Three-language list of botanical name components

Three-language list of botanical name components

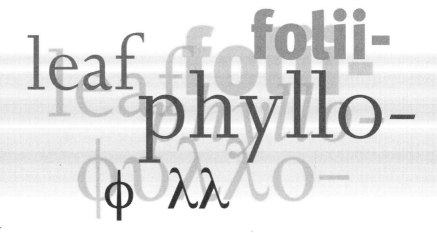

A. Radcliffe-Smith

Published by Royal Botanic Gardens, Kew 1998

CONTENTS

THREE-LANGUAGE LIST OF BOTANICAL NAME COMPONENTS

Back in December 1983, Dr. R.K. Brummitt, in issue no. 2 of his and N.P. Taylor's "Nomenclatural Forum", expressed a wish that a list of Greek and Latin roots or elements used in botanical names and epithets be compiled, in order that mixtures of Latin and Greek in new compound-coinages might be avoided. Over the next few years in the succeeding 16 issues, there appeared such a list, in instalments, which I put together, with Greek roots alphabetically arranged in the first column and their Latin and English equivalents in the second and third columns.

Whilst this was appearing, I was asked by P.S. Green whether or not this information could also be presented alphabetically under Latin and English heads as well. At that time I was totally word-processor illiterate, but bore the request in mind when, some years later, I was introduced to the amazing alphabetizing ability of the Spreadsheet.

Also, in the intervening years, I have been continually making additions to the list that appeared in issues 3–18 of "Nomenclatural Forum", as well as compiling a Classical Plant-name Supplement and a Numerical Supplement. The following offering is the result, and it is hoped that it may help to serve an additional purpose, namely to help its users understand the meanings of Latin and Greek compound-coinages already in the literature.

I would, however, issue a caveat regarding its use in the coinage of new compounds: obviously some possible compounds are going to be less acceptable than others for various reasons, not least of which is that of euphony. Thus, Greek generally lends itself to compound-formation much more readily than Latin does, partly but not entirely due to the fact that very often fewer syllables are involved. For example, if in a genus with normally large leaves a small-leaved species were to be discovered, the Latin-based epithet conveying this would be '*parvifolius*' – a 5-syllabled compound, whereas the Greek equivalent would be '*microphyllus*' with only 4 syllables; moreover, if in a genus with normally warty fruits a smooth-fruited species were to be discovered, the Latin-based epithet conveying this would be '*laevifructus*', and the Greek equivalent would be '*lissocarpus*': the same number of syllables occurs in both, but the latter is a much crisper and more easily-pronounced word.

It is often rather difficult to come up with meaningful names for genera in very large families, and for species in very large genera, for the simple reason that undoubtedly the more obvious coinages will already have been made. Consequently, in the lists that follow will be found some rather unusual word-elements which are not based on the normal word, but perhaps on an abstruse dialectal form, sometimes so abstruse as to be a *hapax legomenon* – a word appearing only once in the Classical Literature. An example is "phalaino-", on which the generic name *Phalaenopsis* is based; "phalaina" or "phallaina" normally means "whale" in Greek, but in the Rhodian dialect it means "moth"! It seems clear that the orchid genus must have been named from the latter!

Another caveat which needs to be issued concerns the shift of meanings in Botanical Latin and Greek from classical usage. Thus, for example, the Greek 'chaete' signifies 'loose flowing hair, mane', whereas botanical usage often indicates 'stiff bristle', and 'pelios' originally referred to parts of the human body discolored by extravasated blood, whereas in botanical usage it has come to mean 'black' generally. Consequently, meanings given in classical Latin and Greek lexicons may well not convey the exact shade of nuance sought by the researcher.

In transliterating from the Greek, authors have not always taken account of the initial rough and smooth breathing signs, hence mistakes like 'ecasto-' instead of 'hekasto-' and 'omalo-' instead of 'homalo-' may be encountered. This of course leads to problems of indexing as well as of priority, as when a later author makes an orthographic correction of the spelling used by an earlier one, whilst another might insist that the original spelling should be retained, even if incorrect.

In view of the need to keep costs down, it was decided not to use the Greek Alphabet, even though in some instances certain ambiguities of orthography would have thereby been clarified.

In 1984, Dr I.K. Ferguson drew my attention to a triple-language list of a more general nature which had been produced in the seventeenth century by John Ray, namely his "Dictionariolum Trilingue" (1675) and which had been issued in facsimile by the Ray Society in 1981. The Greek Alphabet is used in that work, but is difficult to read owing to the rather obscure fusions and abbreviations employed.

I am indebted to Mr. Malcolm Young, former Senior Classics Master at Westminster Under School, and currently Latin Mentor to the Herbarium Staff here at Kew, for his patience in going through the main list and the numerical supplement to root out the many errors and inconsistencies that they were found to contain!

I would like to thank Suzy Dickerson of the Editorial Unit of Kew's Information Services Division for all her help in expediting the production of this list, and my especial thanks also go to Margaret Newman of the same Division for all her hard work in preparing the list for publication.

Finally, if there are any botanists using the list who find that there are roots or elements omitted which they feel should have been included, which I am sure many will, I should be most grateful if I could be informed, in the hope that, in the course of time, a General Supplement, or perhaps even a new edition, can be prepared.

– A. Radcliffe-Smith, Herbarium, RBG Kew, February 1998

NB: The use of parentheses ( ) signifies either that no exact equivalent word exists in one or other of the three languages, or else that the equivalent, if it does exist, cannot be used to form compounds.
'L & S' = Liddell & Scott, A Greek-English Lexicon, ed. 6 (1869); 'LSJ' = do., rev. & aug. H.S. Jones (1925).

GREEK	LATIN	ENGLISH
(no Gr. equiv.)	cuculli-	cowl, hood
aageo-	duri-	hard
a-, an-	e-, ex-	without
acampo-	fragili-	brittle
acantho-	spini-	spine
-achne	lanugini-	removeable covering
achyro-	palea-	chaff
aci-, acido-	acumini-, cuspidi-	point
aci-, acido-	mucroni-, puncti-	point
-acme	acumini-, apici-	point, summit, tip
-acme	cacumini-, summi-	point, summit, tip
acmo-	incudi-	anvil
acrido-	locusta-	grasshopper
acro-	culmini-	top
acro-	summi-	top
actino-	radii-	ray
adelo-	ignoti-	unknown
adelpho-	fratri-	brother
adeno-	glanduli-	gland
adianto-	(no Latin equivalent)	water-repellent
adino-	aggregati-, coacervati-	close, crowded
adino-	coarctati-, conferti-	close, crowded
adino-	congesti-, conglomerati-	close, crowded
adino-	crebri-	close, crowded
aechmo-, aichmo-	hasti-	spear
aechmo-, aichmo-	acumini-, cuspidi-	point
aechmo-, aichmo-	mucroni-, puncti-	point
aego-, aegi-	capra-, capri-	goat
aegilopi-	aveni-	oats
aei-, ai-	semper-	always
aethio-, -eo-	insueti-	unusual
aetho-	ustulati-	burnt
aetio-	causa-	cause
aeto-	aquili-	eagle
aexi-	aucti-	increasing
agamo-	caelibi-	unmarried
agatho-	boni-	good
aglaio-, aglao-	magnifici-	splendour
agnoto-, agnosto-	ignoti-	unknown
agrio-	sylvestri-	wild
agro-	agresti-	field, country
ailuro-	feli-	cat
akaino-	senti-	thorn
akineto-	immobili-	immoveable
akonto-	pili-	javelin
akonto-	jaculi-, teli-	dart
akonto-	hasti-	spear
alectoro-, alectryo-	galli-	cock
aletho-	veri-	truth
aleuro-	farini-	floury, mealy

alkimo-	robusti-, validi-	stout, strong
allago-	mutationi-	change
allanto-	botuli-	sausage
allelo-	mutui-	mutual
allo-	alii-	other
alopeco-	vulpi-	fox
ambly-	obtusi-	blunt
ammo-	areni-	sand
amno-, arno-	agni-	lamb
amoibo-	permutanti-	exchanging
amoibo-	successori-	follower
amphi-	ambi-	on both sides
amphoreo-	dolio-	jar
amydro-	obscuri-	indistinct
amylo-	amyli-	starch
amyxi-	lacerati-	torn, mangled, rent
ana-	re-	again
ana-, ano-	sub-	up
anabaeno-	assurgenti-	rising
anacamps-, anacampt-	reverti-	returning, returned
anatoli-	orienti-	east
-anche	stranguli-	strangle
anchi-	propinqui-	near
ancistro-	hamuli-	fish-hook
ancyclo-	anfractuosi-	crooked
andro-, anthropo-	homini-, viri-	man
aneilemato-	involuti-	rolled up
anemo-	venti-	wind
aneto-, aneuro-	laxi-, remissi-	slack
angelo-	nuntii-	messenger
angio-, ango-	vasi-	vessel
angkale-	flectibrachii-	bent arm
angkalido-	manipuli-	armful
angkalo-	fasci-	bundle
aniso-	inaequi-	unequal
ankono-	flexu-	bend
ankylo-	pravi-	bent, crooked
ankyro-	ancora-	anchor
anomalo-	enormi-	abnormal, irregular
anomalo-	inaequi-, iniqui-	uneven
anophero-	ascendenti-, acclivi-	ascending
anoplo-	inermi-	unarmed
-antha, -e	-flora	-flowered
antheio-	florei-, floridi-	flowery
antheli-	paniculi-	reed-plume
antherico-	spici-	corn-ear
antho-, -anthemon	flori-	flower
antro-	antri-, cavi-	cave
antro-	caverni-, specu-, spelunci-	cave
aphano-	invisi-	unseen
aphelo-	simplici-	plain, simple
apo-	ab-	away from

apobathro-	scala-	ladder
apomono-	solitudini-	isolation
aptosi-, aptoto-	firmi-, stabili-	holding fast to
arachno-	aranei-	spider
araio-	rari-	thin
araro-	conferti-	close-packed
archaeo-, archaio-	veteri-	old, ancient
arche-, archi-	principi-	chief
arcto-	ursi-	bear
arcy-	cassi-	net
argi-, argo-	albi-	white
argillo-	argilla-	clay
argyro-	argenti-	silver
aristero-	sinistri-	left
aristo-	optimi-	best
arma-, arme-	coniuncti-	junction, union
armo-	commissuri-	joint
aromato-	aromati-	spice
arotro-	aratri-, vomeri-	plough
arouro-	arvi-	field
arrheno-, arseno-	mari-, masculi-	male
arthro-	arti-, articuli-	joint
arti-	(iam, nunc)	now
artio-	aequi-, pari-	even
arto-	pani-	bread
arytero-, arytaino-	trulli-	ladle
asco-, cysti-	vesici-, vesiculi-	bladder
asemanto-	exigui-	insignificant
asemno-	ignobili-	undignified
aspalaco-	talpi-	mole
aspasio-	amplexi-	embrace
aspazo-	amplecti-	clinging
aspido-, -aspis	clipea-	round shield
aspilo-, aspiloto-	casti-, immaculati-, puri-	spotless
astaco-	astaci-	lobster
astero-	stellati-	-starred, starry
astheno-	aegri-, morbidi-	diseased
asti-, asty-	urbi-	city
astrapo-	fulguri-	lightning
astro-	stelli-	star
asystasio-, asystato-	confusi-	confused, disordered
atalo-	teneri-	delicate, tender
ateramno-	amari-	bitter
ateramno-	asperi-	harsh
ateramno-	duri-	hard
athanasio-, athanato-	aeterni-, immortali	everlasting, undying
athanasio-, athanato-	sempiterni-	everlasting, undying
athero-	spici-	corn-ear, spike
athro-	conferti-	crowded
atimeto-	contempti-, despicati-	despised
atimeto-	humili-, spreti-	despised
atracto-	fusi-	spindle

5

atropo-	rigidi-	inflexible, unbending
atropo-	inculti-	untilled
atyloto-	indurati-	hardened
auge-	luci-	light
aulaco-	sulcati-	furrowed
-aulax	sulci-, -sulcus	furrow
aulo-	tibii-	flute
aura-	aura-	breeze
auto-	ipsi-	self
auxi-	creti-	growth
axi-, axio-	digni-	worth
axino-	securi-	axe
axono-	axi-	axle
azaleo-	aridi-, sicci-	dry
bactero-, bactro-	baculi-, bacilli-	cane, club, cudgel, staff
baeo-	exigui-, minuti-	little, small
baeo-	parvi-, pusilli-	little, small
balano-	glandi-	acorn
balio-	gutta-, macula-	spot
bapho-, bapsi-	tincti-	dyed
bapti-	mersi-	dipped
barathro-	gurgiti-, profundi-, voragini-	abyss, gulf
barathro-	chasmati-, fissuri-	chasm, cleft
barathro-	hiati-, rimi-	chasm, cleft
barathro-	fovei-, putei-	pit
bary-	gravi-	heavy
basileo-	regi-	king
-bates	-tegens	covering
bathmo-	gradi-, scali-	step, stair
bathmo-	limini-	threshold
bathro-	basi-	base, pedestal
bathy-	alti-, profundi-	deep
batracho-	rani-	frog
bdallo-	mulcti-, mulgi-, mulsi-	milking
bdello-, -bdella	hirudini-	leech
bdello-, -bdella	lampetro-, muraeni-	lamprey
belo-	sagitti-	arrow
belo-	iaculi-, teli-	dart
-belone	acu-	needle
-bius	-vivus	living
blabe-	laesioni-, vulneri-	injury
blabero-, blabo-	nocenti-, nocivi-, noxii-	baneful, damaging, harmful, hurtful, injurious
blapto-, blaptiko-, -blapton	nocenti-, nocivi-, noxii-	baneful, damaging, harmful, hurtful, injurious
blasto-, -blastos	surculi-	shoot
blemmato-, -blemma	oculi-	eye
blemmato-, -blemma	aspectu-	glance, look
blephari-, blepharido-	cilii-	eyelash
blepharo-, -blepharon	palpebri-	eyelid
bolbo-	bulbi-	bulb

bolo-, -bolos	iactu-	throw
bolokopo-	irpici-	harrow
bombyco-	bombyci-	silkworm
boö-, bou-, bu-	bovi	ox
borboro-	coeni-, caeni-	dirt
borboro-	muci-	slime
boreo-	septentrionali-	north
bostrycho-	cincinni-	curl
bothrio-	fovei-	pit
botryo-, -botrys	racemi-, uvi-	cluster
bouno-	colli-	hill
brachio-	brachii-	arm
brachy-	brevi-	short
brizo-	nutanti-, nutati-	nodding
brochido-	arcuati-	looped
brocho-	laquei-	noose
bromato-, -broma	cibi-, pabuli-	food
bromeso-	fremitu-	crackling, noisy, roaring
bromo-	foetidi-	stinking
bronto-	tonitru-	thunder
brosimo-	eduli-, esculenti-	eatable
broto-	homini-, viri-	man
broto-	carni-	flesh, meat
bryche-, brycheto-	fremitu-, mugitu-, strepitu-	bellowing, roaring
bryo-	musc-	moss
butyro-	butyri-	butter
byrso-	aluta-, cori-, pelli-	leather
bysmo-, bystro-	obturaculi-, obturamenti-	bung, plug, stopper
bysmo-, bystro-	obturatori-	bung, plug, stopper
bysso-, bytho-	alti-, altitudini-, ponti-	deep (the), depth
bysso-, bytho-	profundi-	deep (the), depth
cachleco-	calculi-	pebble
caco-	male-	bad
caeno-, ceno-	novi-	new
calatho-, -calathus	corbi-	basket
calche-	murici-	purple
calli-, calo-	pulchri-, venusti-	beautiful, charming, lovely
callyntro-	penicilli-	brush
calpi-	situli-	bucket, urn, vessel
calyco-	(no L equiv)	flower-cup
-calymma	integumenti-, involucri-	covering, a
-calymma	operculi-	covering, a
calypso-, calypto-, -calyptus	tecti-	covered
calyptro-	veli-	veil
camaco-	pedamini-, pedamenti-	vine-prop
camaro-	fornici-	vault
camelo-	cameli-	camel
camino-	fornaci-	kiln
cammaro-	cammari-, gammari-	lobster
campe-	flexi-	bending, a
campso-, campto-, campylo-	flecti-	bent

canono-	virga-	rod
cantharo-	scarabaei-	beetle
capno-	fumi-	smoke
capro-	apri-	wild boar
carabo-	scarabaei-	stag-beetle
carcharo-	serrati-	jagged
carcino-	cancri-	crab
cardio-	cordi-	heart
carpho-	straminei-	straw
-carpa	-fructa	fruited
carpo-	frugi-	fruit
caryo-, -caryon	nuci-	nut
cassitero-	stanni-	tin
castoro-	fibri-	beaver
cata-	infra-, sub-	below, down
catharo-	casti-, puri-	clean, pure
cathartico-	purgativi-	cleansing, purgative
caulo-, -caulon	cauli-, -caulis	stem
caustico-	urenti-	burning, caustic, corrosive
causto-	cremati-, incensi-	burnt
causto-	tosti-, usti-	burnt
cecidio-	galli-	gall
cecido-	succidi-, sucidi-	juice, sap
cecryphalo-	reticuli-	hairnet
celado-	crepitu-, fremitu-, strepitu-	clamour, din, noise
celaeno-, celaino-	nigri-	black
celeutho-	calli-, semita-, tramiti-	path, track
celeutho-	cursu-, itineri-, via-	road, way
celido-	macula-	stain
celypho-	cortici-, crusta-, folliculi-	husk, rind, shell
celypho-	gluma-, putamini-, testa-	husk, rind, shell
celypho-	siliqua-	pod
ceno-	vacui-	empty
centemato-	stimuli-	goad
centro-, -centron	acumini-	point
centro-, -centron	calcari-	spur
-cephala, -e	-capitata, -ceps	-headed
cephalo-	capiti-	head
cepo-	horti-	garden
cepheno-	fuci-	drone
ceramo-	argilla-, creta-	clay
-ceras	-cornu	horn
cerato-	cornuti-	horned
cerauno-	fulmini-	thunderbolt
cerchno-	asperi-	rough
cercido-	feruli-, paxilli-, virga-	peg, rod
cerco-	cauda-	tail
cercopo-	simia-, simii-	monkey
cerio-	favi-	honeycomb
cero-	cornu-, cornui-	horn
cero-	cera-	wax
cesto-	suti-	stitched

cestroto-	acuti-	pointed
ceto-	ceti-	sea-monster
ceutho-	abditi-, celati-, occulti-	hidden
chaeno-	hianti-	gaping
chaero-	laeti-	delighting in, rejoicing in
chaeto-, -chaete	seti-	bristle
chalaro-	laxi-	- loose
chalazo-	grandini-	hail
chalazo-	pustuli-, tuberculi-	pimple, tubercle
chalco-	aeri-	bronze, copper
chalico-	calculi-	pebble
chamae-, chamelo-	demissi-, humili-	low-growing, on the ground
characo-, -charax	-sudes, -vallus	stake
characto-	notati-	marked
charadro-	torrenti-	mountain-stream
charagmato-	nota-	mark
-charis	venusti-	beauty, grace
charito-	gracili-	graceful
charto-	charta-	paper
chasco-	hianti-	gaping
chasmato-	hiatu-	chasm
chauno-	flaccidi-	flabby
cheilo-	labii-	lip
cheimo-	hiemi-	winter
cheiro-, chero-, chiro-	manu-	hand
-chele	uncini-	claw
chelidono-	hirundini-	swallow
chelono-	testudini-	tortoise
cheno-	anseri-	goose
cherado-	glarea-	gravel, shingle
chero-	erinacei-	hedgehog
cherso-	terra-	dry land
chiasmato-	decussati-	crosswise
-chilus	-labiatus	-lipped
chimaero-	capra-	she-goat
chiono-	nivi-	snow
-chiton	tunica-	tunic
-chlaena	palliati-	cloak
chlamydo-	(no L equiv)	cloak
chloano-	virelli-, viriduli-	greenish
-chloe	gramini-	grass
chloro-	viridi-	green
chnoo-	lanugini-	down
choano-, chono-	infundibuli-	funnel
choero-, choiro-	porci-	pig
choiridio-	porcelli-	piglet
chole-, cholo-	bili-	bile
chorde-, chordo-	chorda-	catgut
-chorion	membrana-	membrane
choro-	regioni-	place, plot, region, space
chorto-, -chortos	foeni-	hay

-chroma, -chroos	-color	colour
chromato-	colorati-	coloured
chrono-	tempori-	time
chroso-, chrozo-	tincti-	dyed, stained, tinged
chryso-	aurei-, aureo-	gold
chthamalo-	humili-	low
chthono-	telluri-, terra-	earth
chylo-, chymo-	succi-	juice, moisture, sap
chytro-	olla-	pot
chytro-	fictili-, figlini-	pottery
ciboto-	arca-, capsa-, cista-, pyxi-	box
cinclido-	cancelli-	lattice, trellis
cio-	curculioni-	weevil
ciono-	columni-	pillar
circo-	falconi-	falcon, harrier, hawk
cirrho-, cirro-	fulvi-	tawny
cisto-	cista-	box, chest
cithara-	lyra-	harp, lute, lyre
citrino-	citrini-	lemon-yellow
cladeuto-	putationi-	pruning
cladio-, cladisco-	ramuli-	branchlets
clado-	rami-	branch
clao-	fragili-	brittle
clasmato-	frusti-	fragment, piece
clasto-	fracti-	broken
cleido-	clavi-	key
cleisto-	clausi-	closed, shut
cleithro-	obici-, pessuli-	bolt
cleithro-	repaguli-, sera-	bolt
clemato-, -clema	insiti-, surculi-	graft, scion, slip, twig
-cles	celebri-	celebrated
climaco-, -climax	scala-	ladder
clino-, -cline	cubili-	couch
clono-	surculi-	shoot
clostero-	fusi-	spindle
cneme-	tibii-	shin
cnemido-, -cnemis	ocrei-	legging
cnodaco-	cardini-	pivot
coccino-	coccinei-	scarlet
coccy-, -coccos	bacca-, -coccus	berry
coccygo-	cuculi-	cuckoo
coccymelo-	pruni-	damson, plum
cochlio-	cochlea-	snail
codono-, -codon	campani-	bell
coeleo, coelia-, coelio-	uteri-	womb
coelo-	excavati-, -cavus	hollow
coeno-	communi-, vulgari-	common
colaco-	parasiti-	sponger
coleo-	vagini-	sheath
colla-, collo-	glutini-	glue
colo-	artu-, membri-	limb
colobathro-	gralli-	stilt

colobo-	curti-	curtailed, shortened, stunted
colono-	clivi-, colli-	hill
colopho-	apici-	top
colousto-	truncati-	cut short, docked
colpo-	sinu-	bosom
colymbi-	natanti-	swimming
colymbi-	mersi-	diving
come-, como-, -come	capilli-, crini-, pili-	hair
compso-	eleganti-	neat
concho-	concha-	mollusc, shell
condylo-	articuli-	knuckle
coni-, conio-, conis-	pulveri-	dust
cono-	coni-	cone
conopso-	culici-	gnat
conto-	brevi-	short
cope-	remi-	oar
copho-	obtusi-	blunt
copido-	dolabri-	chopper, cleaver
copido-	falculi-	billhook
copro-	stercori-	dung
copto-	contusi-	beaten, bruised, pounded, struck
coraco-	corvi-	raven
cordyle-	clavi-	club
cordyle-	baculi-, fusti-	cudgel
corema-	scopae-	besom, broom
corethro-	scopae-	besom, broom
corio-	cimici-	bedbug
cormo-	stirpi-	treetrunk
corone-	cornici-	crow
coronido-	curv(at)i-, pandi-, pravi-	bent, crooked, curved
coryco-	culei-	leathern sack
corydo-	alauda-	lark
corymbo-	culmini-, summi-	top
coryne-, coryno-, -coryne	clavi-	club
corypho-	culmini-, summi-	top
corysto-	coacervati-, congesti-	heaped up
corysto-	cumulati-	heaped up
corytho-	crist(at)i-	crest
corytho-	cassidi-, galea-	helmet
coscino-	cribelli-, cribri-	sieve
cosmo-	mundi-	world, universe
cosymbo-	fimbria-	fringe
cotylo-, -cotyle	cupuli-	cup
craspedo-	margini-	border
cratero-, crato-	robusti-, validi-	strong
cratero-	crateri-	bowl
crea-, creio-, creo-	carni-	flesh
cremasto-	pendenti-, penduli-	hanging
cremno-	scopuli-	cliff
crepido-	calceoli-, crepidi-, solei-	sandal, slipper

11

crepido-	calcei-, caliga-, cothurni-	boot, shoe
crico-	annuli-, circuli-	ring
crio-	arieti-	ram
crobylo-	crista-	crest, tuft
crocoto-	crocei-	saffron-coloured
crossoto-	fimbriati-	fringed
crotalo-	crepitu-	rattle
cryo-, crymo-	glaciei-	ice
cryo-, crymo-	gelu-	frost
crypsi-, crypso-	abditi-, arcani-, occulti-	secret
crypsi-, crypso-	remoti-, secreti-	secret
crystallo-	glaciei-	ice
crypto-	abditi-, celati-, occulti-	hidden
crypto-	tecti-	covered
cteno-	pectini-	comb
ctono-, -ctonos	neci-, -cidus	murder
cubo-	cubi-	cube
cyano-	caerulei-	dark blue
cyatho-	acetabuli-	cup
cyclo-	annuli-	ring
cycno-	cygni-	swan
cylico-	calici-, poculi-, scyphi-	cup
cylindro-	cylindri-, scutuli-	roller
cyllo-	pravi-	crooked
cymato-	tumidi-	swollen
cymbi-	cymba-	boat
cymbi-	poculi-	small cup
cymo-	fluctu-	billow, swell, wave
cyno-	cani-	dog
cypello-	poculi-, scyphi-	goblet
cypho-	gibbi-	humped, hunched
cypseli-	alveoli-, capsula-	box, container
cyrillio-	urcei-, urceoli-	jug, pitcher
cyrto-	curvati-	curved
cystho-	alvei-, cavi-, lacuna-	hollow
cysti-	vesiculi-	bladder
cyto-	celluli-	cell
cyto-	cavi-	hollow
cyttaro-	cavi-	hollow
dactylo-	digiti-	finger
dako-, -dakos	morsu-	bite
dako-, -dakos	stimulanti-, urenti-	sting
dakryo-	lacrimi-	tear
dasy-	hirsuti-, hirti-, hispidi-, villi-	shaggy
deino-	diri-, terribili-	dreadful
delo-	manifesti-	evident
delphino-	delphini-	dolphin
delphy-, delphyo-	uteri-	womb
delto-	triangulari-	triangular
dendro-	arbori-	tree
depao-, -depas	calici-	chalice

dermato-, dermo-, -derma	cuti-, pelli-, -pellis	skin
-desme	fasci-	bundle
desmo-, desmos	liguli-, lori-	band
deutero-	secundi-	second
dexio-	dextra-	righthand
dia-	per-	through
diaphoro-	differenti-	different
dicha-, dicho-	binati-	paired
dicraeo-	furcati-	forked
dictyo-	reti-	net
didymo-	bini-, gemini-	twin
diphthero-	aluti-, corii-, pelli-	leather
diplo-	ancipiti-, duplici-	double
diptycho-	duplicati-	doubled, folded
disco-	tori-	quoit
disso-, ditto-	duplici-	double
dolicho-	longi-	long
domatio-	cubiculi-	bedchamber
doro-	donati-	gift
dory-	hasti-	spear
-doxa	gloriosi-	glory
draconto-	draconi-	dragon
-dragma	fasci-, manipuli-, mergiti-	sheaf, truss
drapeto-	fugitivi-	runaway, a
drastico-	efficaci-	active, vigorous
drepano-	falci-	sickle
drimy-	acri-	keen, pungent, sharp
-dromos	-currens	running
droso-	rori-	dew
drymo-	dumeti-, fruticeti-, silvuli-	coppice, copse, thicket
drypto-	lacerati-	torn
dys-	male-	bad, ill
earo-	veri-, verni-, vernali-	Spring
ecasto-	(quisque)	each, every
ecballo-	eiecti-	throw out
ecdysi-	exuti-, exuvi-	slough off
echidno-, echi-	viperi-	adder
echino-	erinacei-	hedgehog
ecteino-, ecteno-	extensi-, extenti-	stretched
ecto-, exo-	extra-	outside
edapho-	humi-, soli-	ground
eido-	forma-, specie-	figure, form, shape
eikono-	imagini-	likeness
eilemato-	involucri-	covering
einosi-	tremuli-	quiver
eiro-	lani-	wool
elachisto-	minimi-	smallest
elachy-	brevi-, demissi-, exigui-	low, short, small, tiny
elachy-	humili-, minuti-	low, short, small, tiny
elachy-	parvi-, pusilli-	low, short, small, tiny
elaio-	olei-	oil

13

elapho-	hinnulei-	fawn
elaphro-	levi-	lightweight
elasmo-	bractei-, catilli-	plate, sheet
elasmo-	lamini-, patelli-	plate, sheet
elasso-, elatto-	minori-	less
elaterio-	expulsi-	driven away
electro-	sucini-	amber
eleuthero-	liberi-	free
eligmato-	cincinni-, cirri-	curl
elytro-	operculi-	cover
embritho-	gravi-, ponderosi-	heavy
embryo-	foeti-, inchoati-	unborn, unformed
empedo-	semper-	ever-
empedophyllo-	sempervirenti-	evergreen
enalio-	marini-	sea, of the
enallago-	alterni-	exchange
enantio-	adversi-, contra-	opposite
enarthro-	articulati-	jointed
encephalo-	cerebri-	brain
enchelyo-	anguilla-	eel
endemo-	incolenti-	dwelling in
endo-, ento-	intra-	within
enoplo-	armati-	armed
enoptro-	speculi-	mirror
entero-	intestini-	gut
entomo-	insecti-	insect
enydro-, enygro-	aquatici-, aquatili-	aquatic
eo-	aurori-	dawn
epeiro-	continenti-	mainland
epeteio-, eteio-	annui-	yearly
epi-	super-	upon
episkynio-	supercilii-	eyebrow
erato-	amabili-, amoeni-, venusti-	lovely
ereismato-	adminiculi-, columini-	prop, stay, support
ereismato-	firmamenti-, fulcimenti-	prop, stay, support
ereismato-	pedamenti-	prop, stay, support
eremo-	deserti-	desert
erepsi-	teguli-	roofing
erio-	floccosi-, lanati-	woollen
eripho-	haedi-, hoedi-	kid
erodio-	ardei-	heron
erymno-	septi-	fence
erysibo-	robigini-	mildew
erythro-	rubri-	red
eschato-	ultimi-	last
esoptro-	speculi-	mirror
esotato-	intimi-	innermost
esotero-	interiori-	inner
esthe-	vesti-	clothing
etheiro-	capilli-, crini-	hair
ethmo-	cribri-	colander, sieve, strainer
ethmoideo-	perforati-	perforated

eu-	bene-	well
euod-, euosm-	fragranti-	fragrant
euonymo-	sinistri-	lefthand
eury-	lati-	broad, wide
eusticto-	maculosi-	variegated
euthy-	recti-	straight
execho-	prominenti-	raised
exo-	extra-, ultra-	away from, beyond, out of
exoncomato-, exoncosi-	umboni-	shield-boss, protuberance
galacto-	lacti-	milk
gale-, gali-	musteli-	weasel
gamo-	connati-	united
gano-	nitidi-	bright
gastro-	ventri-	belly
geiso-, geisso-	protecti-, sima-	cornice, eaves
-geiton, -geton, -giton	vicini-	neighbour
genea-	stemmati-	pedigree
geneiado-	barba-	beard
geneio-	menti-	chin
-geneus	-genus	kind, of a
geo-	telluri-, terrestri-	land
gephyro-	ponti-	bridge
geraio-, geronto-	seni-, veteri-	old
gerano-	grui-	crane
ginglymato-	cardinali-	hinged
ginglymo-	cardini-	hinge
glaphyro-	excavati-	hollowed
glaphyro-	politi-	finished, polished, smooth
glauco-	glauci-	blue-grey
glochidio-	aculeoli-	point (dim.)
-glochin	aculei-	point
gloeo-	glutinosi-	glue
glossario-, glossidio-	linguli-	small tongue
glosso-, glotto-	lingui-	tongue
glouto-	cluni-	rump
glyco-	dulci-	sweet
glypho-, glypto-	sculpti-	carved
gnatho-	maxilli-	jaw
gnomono-	indici-	pointer
gompho-	clavi-	nail
gompho-	obici-, pessuli-	bolt
gompho-	repaguli-, sera-	bolt
-gone	genitali-	reproductive organs
gongylo-	nodosi-, torulosi-	knobbly
gongylo-	globosi-, glomerosi-	round
gongylo-	orbiculati-, rotundi-	round
-gonia	anguli-	angle
gonio-	angulati-	angled
gony-	genu-	knee
gony-	geniculati-	-kneed
gorgo-	terribili-	frightful

15

goryto-	pharetra-	quiver
grammato-, -gramma	litteri-, scripti-	letter, writing
-gramme	linei-	line
grapho-, grapto-	delineati-	drawn
grapsaio-	cancri-	crab
gripho-	reti-	fishing-net, -basket
gromphado-	scrofa-	old sow
grypo-	adunci-	hooked
grypo-	gryphi-	griffin
gyalo-	cavi-	hollow, a
gyio-	arto-, membri-	limb
gyno-	foeminei-	female
gyro-	circulari-, rotundati-	round
habro-	teneri-	delicate
hadro-	corpulenti-, crassi-, obesi-	bulky, stout, thick
haemato-	cruenti-, sanguinei-	bloody
halicacabo-	physali-	wintercherry
halo-	sali-, salso-	salt
halysi-	cateni-	chain
hama-	co-, col-, com-, con-, cor-	together with
hapalo-	molli-	soft
hapalo-	teneri-	delicate
hapax-	singuli-	once
haplo-	soli-	single
hapto-	tenenti-	holding
harpago-	hami-, unci-	hook
harpazo-	prehensi-	seizing
hebdoma-	hebdoma-	week
-hedra	sedi-, sedili-	seat
hedraio-	sessili-	seated
hedy-	amoeni-, grati-, iucundi-	agreeable, pleasant
hedy-	dulci-, suavi-	delightful, sweet
hedyosmo-, hedypnoi-	fragranti-	pleasant-scented
hegemono-	duci-	guide
heleo-, helo-	palustri	marsh
helico-	cochlei-	spiral
helicto-	cochleati, torti-	twisted
heligma-	cincinni-, cirri-	curl
helino-	capreoli-, clavicula-	tendril
helio-	soli-	sun
helmintho-	vermi-	worm
hemero-	diurni-	day
hemi-, hemisy-	semi-	half
hepato-	iecori-	liver
herpesti-	repenti-	creeping
herpeto-	angui-, serpenti-	snake
hespero-	occidenti-	west
hespero-	vesperi-	evening
hesychio-	taciti-	quiet
hetero-	dissimili-	different
hieraco-	accipitri-	hawk

16

hiero-	sacri-, sancti-	sacred
himanto-	lori-	strap, thong
himatio-	pallii-	mantle
hippario-	equulei-, equuli-	colt, foal, pony
hippo-	equi-	horse
histio-, histo-	carbasi-, veli-	sail
hodego-	duci-	guide
hodo-	via-	way
holo-	integri-	entire, whole
homalo-	plani-	even, level
homo-	(idem)	same
homoio-	simili-	similar
hople-	unguli-	hoof
hoplo-	scuti-	oblong shield
horaio-	hornotini-, tempestivi-	seasonable
horio-, horismo-	fini-	boundary
horio-	maturi-	ripe
hormo-	monili-	necklace
horo-	hora-	hour
hyalo-	vitri-	glass
hybo-	gibberi-, gibbi-	hump, protuberance
hydato-, hydro-	aqua-	water
hydno-	tuberi-	truffle
hyeto-, -hyet	pluvii-	rain
hygro-	humidi-, made-	moist
hylaeo-, hyleo-, hylo-	sylvestri-	forest
hylisto-	percolati-	filtered, strained
hymeno-	membrani-	thin-skinned
hyno-	vomeri-	ploughshare
hyo-, syo-	sui-	sow
hyparxeo-	essenti-, natura-, vita-	being, essence, existence
hypato-	summi-	uppermost
hyper-	super-, supra-	above, over
hyperphoro-	elevati-	raised
hypo-	sub-	below, under
hypselo-, hypsi-	alti-, elati-, proceri-	high
hypsilo-	(no Latin equivalent)	Y-shaped
hypsisto-	altissimi-	highest
hypsitero-	altiori-	higher
hyptio-	supini-	laid back
hyraco-	sorici-	shrew
hystera-	uteri-	womb
hystero-	serotini-	late
hystricho-	hystrici-	porcupine
ianthino-	violacei-	violet-coloured
iatro-	medic-	doctor
ichno-	calli-	track
ichoro-	seri-, sero-	fluid, juice
icmado-	humidi-, rori-	moisture
ictero-	icteri-	jaundice
ichthyo-	pisci-	fish

-idion	-ellus, -illus, -ulus	(diminutives)
ily-, ilyo-	limi-	mud
inio-	fibra-, nervi-	fibre, sinew
ino-	tori-	muscle
-inodes	fibrati-, nervosi-	fibrous, sinewy
io-	viola-	violet
ios-	veneni-	poison
iphi-	validi-	strong
ipno-	furni-	oven
ischado-	fici-	fig
ischio-	coxi-	hip
ischno-	gracili-	slender
ischyro-	validi-	strong
iso-	aequi-	equal
isthmo-	cervici-	neck
-ites, -itis	-aceous, -ago, -eus	likeness
ithy-	recti-	straight
ityo-	margini-	edge
ixod-	visci-	sticky
keimelio-	coacervati-, collecti-	heaped up, stored up
kemado-	hinnulei-	fawn
kero-	corni-, cornu-	horn
keryko-	nuntii-	herald
klasto-	fracti-	broken
koilo-	cavi-	hollow
koilia-, koilio-	uteri-	womb
kentro-, -kentron	calcari-	goad, spur
labyrintho-	ambagi-, labyrinthi-	maze
lacco-	foramini-, fovea-, putei-	hole, pit
lachno-	lanugini-	down
lagaro-	angusti-	narrow
lagaro-	tenui-	thin
lageno-	ampulla-, lagena-, laguncula-	flagon, flask
lago-	lepori-	hare
lagyno-	ampulla-, lagena-, laguncula-	flagon, flask
lailapo-	procella-, turbini-	hurricane
laimo-	fauci-	throat
lampado-	faci-, funali-, taeda-	torch
lampro-	nitidi-	bright
laparo-	laxi-	slack
larco-	corbi-	basket
larnaco-	arca-	box
laryngo-	fauci-	throat
lasio-	villosi-	shaggy
lathro-	abditi-, arcani-, conditi-	secret
lathro-	occulti-, remoti-	secret
lathro-	secreti-, tecti-	secret
lecano-	crateri-	bowl
lecytho-	ampulli-	flask
leicheno-	scabie-	scurf

-leimon, -limon	prati-	meadow
leio-	laevi-, laevigati-	smooth
lemmato-	gluma-	husk
leno-	alvei-	trough
leonto-	leoni-	lion
lepado-	lepadi-	limpet
lepido-	squami-	fish-scale
lepismato-, lepyro-	cuti-, tunica-	peel, rind
lepto-	tenui-	thin
leuco-	albi-	white
lignyo-	fuligini-	soot
likno-	vanni-	winnowing-fan
limno-	stagnali-	pond
lipa-	olei-	oil
liparo-	clari-	bright
lipo-	(sine)	without
lirio-, leirio-	lili-	lily
lispo-, lisso-	laevi-, laevigati-	smooth
litho-	lapidi-, saxi-	stone
lito-	simplici-	simple
lobo-	lobi-	lobe
lochmaio-	dumeti-	thicket
lomato-, -loma	fimbri-	fringe
loncho-	lancea-	lance, spear
lopado-	catilli-, catini-	dish
lopho-	crista-	crest
lopho-	jugi-	ridge
lopo-	cuti-, tunica-	peel, rind
loxo-	obliqui-	slanting
lychni-, lychno-	lumini-	lamp, light
lyco-	lupi-	wolf
lycophoto-	diluculi-	gloaming
lygo-	vimini-	willow-twig
lynco-	lynci-	lynx
lyro-	lyra-	lyre
lysi-	solventi-	loosing
machairo-	acinaci-	sabre
macro-	longi-	long
macryno-	remoti-	distant
malaco-	leni-, miti-, molli-	soft
mallo-	lani-, velleri-	wool
maranto-	deflorescenti-, marcescenti-	fading, withering
maranto-	putrescenti-	decaying
margarito-	margarita-	pearl
marmaro-	marmori-	marble
marro-	bipalii-	mattock
maschalo-	axilli-	armpit
mastigo-	flagelli-	whip
masto-	mammi-	breast
mega-	magni-	great
megalo-	grandi-, grossi-	large

19

megisto-	maximi-	very big
meio-	minori-	less, smaller
meizo-	maiori-	greater
melancho-, melano-	nigri-	shiny black
melancho-, melano-	atri-	dull black
meleagrido-	meleagridi-	guineafowl
meli-	melli-	honey
melino-	lutei-	quince-yellow
melisso-, melitto-	api-	bee
melo-	ovi-	sheep
melo-	mali-	apple
meni-	luna-	moon
meningo-	membrani-	membrane
menoidi-	lunati-	crescentic
mero-	femori-	thigh
-meros	parti-	part
mesembria-	meridiei-	noon
meso-	medi-	middle
meta-	iuxta-	next to
meta-	inter-	among
meta-	mutati-	changed
meta-	post-	after
meteoro-	aeri-	air
metopo-	fronti-	forehead
metoporino-	autumnali-	autumnal
metre-, metrio-, metro-	mensura-	measure
metro-	matri-	mother
micro-	parvi-	small
microtato-	minimi-	least
microtero-	minori-	lesser
milto-	minii-	vermilion
mito-	fili-	thread
mitro-	mitri-	headdress
mixo-	mixti-	mixed
molybdo-	plumbi-	lead
mono-	uni-	single
morpho-, -morpha	formi-, -formis	shape
moscho-	vituli-	calf
myako-	mitylli-	mussel
myceto-, myco-, -myces	fungi-	fungus
mycho-	recessi-	nook
myctero-	nasi-	nose
myelo-	medulla-	marrow
myio-, -myia	musci-	fly
mylo-	mola-	mill
myo-, -mys	muri-	mouse
myo-, -mys	musculi-	muscle
myrio-	innumeri-	countless
myristico-	fragranti-	fragrant
myrmeco-	formica-	ant
myrmedono-	grumuli-	ant's-nest
myro-	odori-	perfume

myro-	collyrii-, unguenti-	ointment
mystaco-	(no Latin equivalent)	moustache
mystro-	cochleari-	spoon
myxo-	muci-	slime
naco-	velli-	fleece
nano-	pumili-	dwarf
nebro-	hinnulei-	fawn
necro-	morti-	dead
necto-	natanti-	swimming
nemato-	fili-	thread
nemeo-	nemori-	glade
neo-	novi-, novo-	new
nepheo-, nepho-, nephelio-	nubi-	cloud
nephro-	reni-	kidney
nesido-	(insula parva)	islet
neso-	insuli-	island
nesso-, netto-	anati-	duck
neuro-	nervi-	fibre, ligament, nerve, sinew, tendon
niphado-	nivi-	snow
nomo-, -noma	pascui-	pasture
notero-	humidi-, made-	damp, moist, wet
notho-	nothi-	bastard
noto-	australi-, meridionali-	south
noto-	dorsi-, tergi-	back
nycti-, nycto-	nocti-	night
nycteri-	vespertili-, vesperugini-	bat
obelo-	veru-	spit
ocheto-	tubi-	pipe
ochlo-	turba-	crowd
ochro-	ochracei-, sili-	ochreous, pale yellow
ochtho-	aggeri-, tumuli-	bank, mound, hillock
ochyro-	firmi-, solidi-, stabili-	firm, lasting, stout
ocy-	celeri-, citi-, rapidi-, veloci-	swift
odonto-	denti-	tooth
-oecium(-oikion)	aedi-	house
oedo-	tumidi-	swollen
oeno-	vini-	wine
-oides	simili-	like
oio-	soli-	alone
oistro-	asili-, tabani-	gadfly
olene-, oleno-, olecrano-	cubiti-	elbow
oligo-	pauci-	few
ombro-	imbri-	rain, shower
ommato-, -omma	ocelli-	eyelet
omo-	humeri-	shoulder
omoplato-	scapuli-	shoulder-blade
omphaco-	immaturi-	unripe
omphalo-	umbilici-	navel
onco-, -oncodes	strumosi-	bulky, swollen

21

Three-language list of botanical name components

ono-	asini-	ass
onomato-	nomini-	name
ontho-	fimi-	dung
onto-	essenti-, natura-, vita	being, essence, existence
onycho-	ungui-	claw
oo-	ovo-	egg
ophio-	serpenti-	snake
ophryo-, -ophrys	supercili-	eyebrow
ophthalmo-	oculi-	eye
opistato-	postremi-	last
opistho-	posteri-	behind, back
opo-	succi-	juice, sap
-ops	personati-	face
opse-, opsi-	serotini-	late
-opsis	faciei-	appearance
orchi-	testi-	testicle
oreo-, ores-, oro-	monti-	mountain
ormeno-	surculi-	shoot
orneo-, ornitho-	avi-	bird
orophe-	tecti-	roof
orphno-	fusci-	dusky
ortho-	recti-	straight
orthrio-, orthro-	aurora-, diluculi-	dawn
ortygo-	coturnici-	quail
orycho-	fodi-	burrow, dig
orycto-	fossi-	burrowed, dug
orygo-	dolabri-	pickaxe
oscho-, -osche	surculi-	scion, shoot, twig
osm-, osmo-, -osme	odori-, olor-	smell
osprio-	legumini-	pulse
osteo-, osto-	ossi-	bone
ostraco-	crusta-	shell
-osyne, -otes	-ugo	(substantival suffixes attached to adjectives)
oteilo-	cicatrici-	scar
othono-, othonio-	lini-, lintei-	flax, linen
otio-	auriculi-	little ear
oto-	auri-	ear
oxy-	acri-	sharp
oxy-	acuti-	acute
ozo-	foetidi-, graveolenti-	stinking
pachne-	pruina-	hoar-frost, rime
pachy-	crassi-, spissi-	thick
-paegma	ludibri-	plaything
pageto-	gelu-, geli-	frost
palaeo-	veteri-, vetus-	old
palin-	re-	again, back
pan-, panto-	omni-	all
papyro-	charta-	paper
para-	iuxta-	beside
paralia-	maritimi-	seaside, by the

partheno-	virgini-	maiden
-patane	-patina, -patella	flat dish, patella
patho-	passi-	suffering
pauro-	parvi-	little, small
pauro-	pauci-	few
pecheo-, pechy-	cubiti-, ulni-	forearm
pedalio-	gubernaculi-	rudder
pedilo-	calce-	sandal, shoe
pedo-	soli-	soil
pegmato-, -pegia	-fixus, -fixa	fixed, made fast, secured
pego-	fonti-, scaturigini-	spring, a
pelago-, pelagio-	marini-	sea, of the
pelargo-	ciconii-	stork
peleceo-, pelecy-, pelyco-	securi-	axe
pelio-, pelidno-	atro-	black, dark
pelio-, pelidno-	lividi-	leaden
-pelma	-firmamentum	prop, stalk, support
pelmato-	planta-, planti-	sole of the foot
pelo-	luti-	clay
peloro-	immani-, ingenti-, monstrosi-	huge, monstrous, prodigious, terrible
pelto-	clipeo-	small round shield
pelto-	parma-	target
pemphigo-	bullati-	blistered, blown out, bubbled
pepeiro-	maturi-	ripe
pepono-	miti-	mellow
pera-	marsupi-, sacco-	pouch
peran-	trans-	across
percno-	fusci-	dusky
perdico-	perdici-	partridge
peri-	circum-	around
periallocaulo-	scandenti-	climbing
perisso-	ampli-	ample
perisso-	copiosi-	abundant, copious
perisso-	immani-	prodigious
perisso-	impari-	odd
perisso-	inaequi-	uneven
perisso-	insigni-, praecipui-	remarkable
perisso-	nimii-	excessive
perisso-	rari-	uncommon
pero-	manci-	maimed
perono-, -perone	acu-	pin
perysino-	annotini-	last year's
petalo-	plani-	flat
petaso-	petasi-	hat
petro-	rupi-, saxi-	rock
phacelo-	fasci-, fasciculi-	bundle, fagot
phaco-, phako-	lenti-	lentil
phaedro-, phaeno-	nitenti-	shining
phaeo-	fusci-	dark, dusky
phago-	esu-	eating

phalacro-	calvi-, glabri-	bald
phalaino-	tinei-	moth
phalaro-	fulici-	coot
phanero-	manifesti-	evident
phano-	nitenti-	shining
pharango-	fissura-, hiatu-, rima-	chasm, cleft, gully, ravine
pharmaco-, pharmako-	veneni-	poison
pharyngo-	fauci-	throat
phascolo-	sacci-	bag
phaulo-	male-, mali-	bad, evil
phello-	suberi-	cork
phenaco-, -phenax	fraudi-	cheat, impostor
phengo-	luci-	light
phere-	ferenti-	bringing
phiali-, phialo-	poculi-	winecup
philo-, -philus	amanti-	loving
phlebo-, -phleps	vena-	vein
phlegmato-	pituita-	mucus
phloeo-, phloio-	cortici-	bark
phlogi-, -phlox	igni-, flamma-	flame
phlyctaino-	pustuli-, vesici-	blister, pustule, vesicle
phobero-	formidabili-	fear-generating
phoeniceo-, phoenico-	punicei-	carmine, crimson
pholido-, -pholis	cornei-	horny reptile-scale
phono-	neca-, -cidus	murder
phono-, -phone	soni-	sound
phorino-	corii-, pelli-	hide
phormo-	storei-, tegeti-	mat
-phoros	-fer, -ger	bearing, carrying
phostero-	astri-, stelli-	star
photo-	luci-	light
phoxi-	fastigiati-	tapering
phragmo-, phragmato-	septi-	fence, hedge, partition
-phragma	septi-	fence, hedge, partition
phrisso-, phrixo-	horridi-	bristling
phrygano-	ligni-	stick, wood
phryno-	bufoni-	toad
phtheiro-	pediculi-	louse
phthinoporo-	autumnali-	autumn
-phthora	putridi-	decay
phyco-	algi-	seaweed
-phye(s)	auctu-, incrementi-	growth
phylaco-, -phylax	custodii-	guard
-phylla	-folia	-leaved
phyllo-	folii-	leaf
phylo-	classi-	class
phylo-	ordini-	order
phylo-	tribu-	tribe
phylo-	geni-	race
phymato-	verruci-	wart
physo-	folli-	bellows
physeto-	inflati-	blown out

phyto-	planti-	plant
piaro-	adipi-, pingui-	fat, grease
picro-	amari-	bitter
pileo-, pilo-	coacti-	felt
pimeleo-, pimelo-	adipi-, pingui-	fat, grease
pinaco-, -pinax	tabuli-	tablet
pinaro-	sordidi-	dirty
pipto-	caduci-	fall
pisso-	pici-	pitch
pitheco-	simii-	monkey
pitho-	dolii-	jar
pityro-	furfuri-	bran
placo-	patelli-	plate
placo-	tabuli-	slab, tablet
placo-	plani-	flat
plagio-	obliqui-	slanting
platy-	lati-	broad, wide
pleco-, plecto-	intexti-, plicato-	plaited, woven
plectro-	inflicti-	strike
plectro-	calcari-	goad, quill, spur, sting
plegmato-, -plegma, plexi-	torti-	plaiting, twisting, weaving
pleio-	pluri-	more
pleisto-	plurimi-	most, very many
pleo-, plero-	pleni-	full
plesio-	proximi-, vicini-	near, neighbour
pletho-	multi-	much
pleuro-	lateri-	side
pleuro-	costa-	rib
plintho-	lateri-	brick
-ploca, -ploce	cincinni-, cirri-	curl
-ploca, -ploce	textili-	web, woven
plocio-, ploco-	cincinni-, cirri-	curl
plocio-, ploco-	textili-	web, woven
pneumatico-	inflati-	blown out
pneumato-	venti-	wind
pneumato-	anima-	spirit
pneumono-	pulmoni-	lung
pnicto-	suffocati-	choked
poco-	velleri-	fleece
podo-, -poda	pedi-, -pes	foot, -footed
poecilo-, poikilo-	variegati-	many-coloured
pogon(o)-, -pogon	barba-	beard
-poiea, -poiesis, -poios	-ficiens	making
-poietos	-factus	made
polio-	cani-, incani-	grey, grizzled
polo-	cardini-	pivot
poly-	multi-	many
pomato-, pomatio-	operculi-	cover
ponto-	mari-	sea
-poros	ori-	opening
porphyreo-, porphyro-	purpureo-	purple
potamo-	amni-, flumini-, fluvii-	river

poterio-	poculi-	winecup
prasino-	porracei-	leek-green
prau-	leni-, miti-, molli-	mild, soft
premno-	caudici-, stipiti-	stump, trunk
prene-	proni-	lying forwards
priono-	serri-	saw
prismato-, pristo-	serrati-	cut
proio-	matutini-	morning
proio-	praecoci-	early
proikto-	praediti-	endowed
pro-, proso-	prae-	before, in front of
pros-	iuxta-	near, close by
protero-	praecociori-	earlier
proto-	primi-	first
prumno-, prymno-	postremi-, ultimi-	last
psacado-	frustuli-, mica-	crumb, morsel
psalido-	forfici-	shears
psammo-	arena-, sabuli-	sand
psaro-	guttati-, maculosi-	dappled
psathyro-	fragili-, friabili-	brittle, crumbly, loose
psedno-	calvi-, glabri-	bald
psepho-	calculi-	pebble
pseudo-	falsi-, spurii-	false, spurious
psicho-	frustuli-, mica-	crumb, morsel
psilo-	calvi-, glabri-	bare
psithyro-	susurri-	whispering
psittacin-	psittaci-	parrot
psocho-	pulveri-	dust
psom(i)o-	frustuli-, mica-	crumb, morsel
psopho-	strepiti-	noisy
psoro-	scabri-	rough, scabby
psycho-	animi-	breath, life, soul
psychro-	frigidi-	cold
psydraco-	pustuli-, vesici-, vesiculi-	blister, pustule, vesicle
psyllo-	pulici-	flea
ptaco-, ptoco-	lepori-	hare
ptairo-, ptaero-	sternuti-	sneeze
pteno-	pennati-, plumati-, plumosi-	feathered
pteri-, pterido-	filici-	fern
pteridio-, pterino-	pennati-, plumati-, plumosi-	feathered
pterno-, -pterna	calci-	heel
ptero-	alati-	winged
pterygio-	fastigi-	gable, pinnacle, roof
pterygo-, -pteryx	ala-	wing
ptilo-	penni-, plumi-	feather
ptortho-	surculi-	shoot, sucker
ptoseo-, -ptosis	caduci-	fall
ptoto-	casi-	fallen
ptycho-	plicati-	folded
ptygmato-	plicati-	folded
ptykto-	plicati-	folded
ptyo-	vanni-	winnowing-fan, -shovel

ptysi-	spuenti-	spitting, a
ptysmato-	sputi-	spittle
ptysso-	plica-	fold
ptyxo-	plicanti-	folding, a
-pus, -podos	-pes	foot
pycno-	densi-	dense
pygo-	cluni-	rump
pylo-, -pyle	-foris, -porta	gate
pymato-	postremi-, ultimi-	last
pymato-	extremi-	outermost
pyndaco-	fundi-	base, bottom
pyreno-	putamini-	fruitstone
pyrgo-	turri-	tower
pyro-	flagr-, flammi-, igni-, incensi-	fire
pyro-	tritici-	wheat
pyrrho-	rubidi-	deep red
pyrrho-	rufi-	orange-red
pyrso-	flammei-, ignei-	fiery red
pythmeno-	caudici-	stock
pythmeno-	fundi-	base, bottom
pyxido-, -pyxis	capsi-	box
regmato-	rima-	break, fracture, laceration, rent
regmato-	rima-	chasm, chink, cleft
rhabdo-	bacilli-, virga-	rod
rhabdo-	sera-	bar
rhabdoto-	vittati-	striped
rhachi-	spina-	backbone
rhacoi-, -rhacodes	lacerati-	ragged, torn
rhadamno-	virga-	shoot
rhadico-	rami-	branch
rhagado-	rima-	chink, crack, rent
rhago-	bacci-	berry
rhago-	acini-	grape
rhaibo-	flexi-, obstipi-	bent, crooked
rhammato-	suti-	sewn
rhammato-	fili-	thread
rhamphi-, rhampho-	rostri-	beak
rhaphi-	suturi-	seam
rhaphido-	acu-	needle
rhapido-	bacilli-, virga-	rod
rhapso-, rhapto-	suti-	sewn, stitched
rheithro-	flumini-	stream
rheo-	fluxi-, flucti-	flow
rhetino-	resina-, resini-	resin
rheumato-	fluxi-, flucti-	flow
rhexi-	fissi-	cracked, split
rhigo-, rhigio-	gelu-	frost
rhikno-	rugosi-	shrivelled
rhino-	nasi-	nose
rhipi-, rhipidi-	flabelli-	fan

27

rhipso-	iactu-	throw
rhipto-	iacti-	thrown
rhizo-, -rrhiza	radici-	root
rhodo-	roseo-, rosi-	rose
rhoiko-	flexi-, flecti-	bent
rhombo-	turbini-	top
rhomphaio-	gladii-	broadsword
rhopalo-	clavi-	club
rhyaco-	torrenti-	stream
rhymbo-	voluti-	coiled
rhyncho-	rostri-	beak, snout
rhyparo-	sordidi-	dirty, filthy, shabby, soiled
rhyso-, rhysso-	rugosi-	wrinkled
rhyti-, rhytido-	rugi-	wrinkle
-rrhoe, -rrhoea, -rrhoa	fluxi-, flucti-	flow
sacco-	sacci-	bag, sack
sacto-	farcti-	stuffed
sagmato-	sella-	saddle
salpingo-, -salpinx	buccina-	bugle, trumpet
sanido-	tabuli-	board, plank
sapro-	putridi-	rotten
sarco-	carni-	flesh
sarmato-	hiatu-	chasm
saro-	versi-	sweep
saro-	scopi-	broom
sathro-	putridi-	rotten
sauro-	lacerti-	lizard
scaio-	scaevi-	lefthand
scaleno-	inaequi-	unequal
scalido-	sarculi-	hoe
scalopo-	talpi-	mole
scambo-	flexi-	bent, crooked
scapano-, -scapane	pala-	shovel
scaphido-	crateri-	bowl
scapho-	alvei-	hollowed
scapho-	lintri-, naviculi-	boat
scapo-	scapi-	shaft, stem
scapto-	fossili-	dug
scato-	fimi-	dung
scedasto-	sparsi-	scattered
scelo-	cruri-	leg
scepano-	tectati-	covered
-scepasma	tecti-	shelter
sceptro-	baculi-, fusti-, scipioni-	staff
schastero-	tendicula-	trap, part of a
schede-	tabula-	tablet
schedo-	comini-, propinqui-	near, nigh
schemato-	habitu-	figure, form, shape
schidio-	assuli-, scinduli-	splinter
schis-, schismato-, schisto-	fissi-, scissi-	split
schizo-	fissi-, scissi-	split

schoeno-	funi-, resti-	rope
schyro-	erinacei-	hedgehog
scia-, scio-	umbra-	shade, shadow
sciado-	umbelli-	parasol
scioto-, -sciodes	umbrati-	shaded
scirrho-	crusta-	rind
sciuro-	sciuri-	squirrel
sclero-	duri-	hard
scoleco-	lumbrici-	earthworm
scolio-	flexi-	bent
scolo-, scolopo-	pali-	stake
scopelo-	scopuli-	rock
scorodo-	allii-	garlic
scorpio-	scorpii-	scorpion
scoto-	tenebri-	darkness
scybalo-	stercor-	dung
scylaco-	catelli-	pup
scypho-	cupuli-	cup
scytalo-	fusti-	cudgel
scyto-	aluta-, cori-, pelli-	leather
seleni-	luna-	moon
-sema	signi-	mark
semeio-, -semeia	vexilli-	standard
serico-	sericei-	silk
siagono-	maxilli-	jawbone
sidero-	ferri-	iron
sigmato-, sigmoideo-	(no Latin equivalent)	S-shaped
simo-	simi-	snub-nosed
sindono-	byssi-	fine linen, muslin
siphono-, -siphon	tubi-	pipe
-sira, -seira	funi-, resti-	rope
siro-	siri-	pit
sisyro-	hircipelli-	goatskin
sitio-	cibi-, pabuli-	food
sitio-	pani-	bread
sito-	frumenti-	corn
smaragdo-	smaragdi-	emerald
smegmato-	saponi-	soap
smyrno-	murra-	myrrh
soleno-, -solen	tubi-	pipe
somato-	corpori-	body
sompho-	fungi-	sponge
soro-	cumuli-, grumuli-, tumuli-	heap, mound
sotro-	curvatura-	felloe
spadico-	spadicei-	brown
spadico-	rami-	branch
spalaco-	talpi-	mole
spano-	pauci-	few
spanisto-	rari-	scanty
sparasso-, sparatto-	lacerati-	torn
spargano-	incunabuli-	swathed
sparto-	funi-	cable

spatho-	lamina-	blade
spato-	aluta-, cori-, pelli-	hide, leather, skin
speiro-	voluti-	coil
spelaio-	caverni-	cave
-sperma	-semina	-seeded
spermato-	semini-	seed
sphaero-	globi-	ball
sphalmato-	lapsu-	fall, slip
spharago-	strepitu-	noisy
spheco-	vespa-	wasp
spheno-	cunei-	wedge
sphigmato-	ligati-, stricti-, vincti-	bound
sphinctero-	ligamenti-	band
sphincto-	ligati-, stricti-, vincti-	bound
sphinxi-	constrictioni-	constriction
sphondylo-	vertebra-	backbone
sphragido-	sigilli-, signi-	seal, stamp
sphragisto-	sigillati-, signati-	sealed, stamped
sphyra-	mallei-	hammer
sphyro-	tali-	ankle
spilo-	macula-	spot, stain
spithami-	dodranti-	span
-spiza	-fringilla	-finch
splanchno-	viscera-	bowels, entrails, gut
spleno-	lieni-	spleen
spodio-	cinerei-	ash-grey
spodo-	cineri-	ashes
spongio-, spongo-	fungi-	sponge
sporadico-	sparsi-	scattered
sporo-	semini-	seed
spyrido-	sporti-	basket
stachyo-, -stachys	spici-, -spica	ear of corn, spike
stacto-	gutta-	exudate, ooze
stagono-	gutta-	drop
stalagmato-	stillicidii-	drip, drop
stalico-, stalido-	pali-	stake
stamno-	olla-	jar
staphido-	astaphi-	raisin
staphylo-	uva-	bunch of grapes
staphylo-	acini-	grape
stathmo-	stabuli-, tegmini-	stable, stall, shelter
stauro-	cruci-	cross
stearo-, steato-	sebi-	suet, tallow
stegano-, -stegia,	tecti-	shelter
stegno-, stego-	tecti-	shelter
steira-	carini-	keel
steiro-	sterili-	barren
stelecho-	caudici-	stock
steleo-	ansi-, capuli-, manubri-	handle
stelo-, -stele	columni-	pillar
-stelma	serti-	crown, garland, wreath
stelmono-	cinguli-, zoni-	belt, girdle

-stemma	corona-	crown
-stemma	serti-	garland, wreath
stemono-, -stemon	stamini-, -stamineus	stamen, thread
steno-	angusti-	narrow
stephano-, -stephus	corona-	crown
stephano-, -stephus	serti-	garland, wreath
stereo-	solidi-	firm, solid
sterigmo-	fulti-	prop, support
steripho-	solidi-	firm, solid
sterno-	pectori-	breast
sterrho-	validi-	strong
stetho-	pectori-	chest
stheno-	robusti-, validi-, viri-	strength
stibado-	culciti-	mattress
sticho-, -stichos	-farius, ordini-, serie-	rank, row
sticto-, -sticta	guttati-, maculati-, punctati-	spotted
-stigma	-nota	mark
stigmato-	notati-	marked
stilbo-, stilpno-, stilpsi-	nitenti-, splendenti-	glossy, polished, shining
stipho-, stiphro-, stipto-	compacti-	close-pressed
stizo-	guttati-, maculati-, punctati-	spotted
stizo-	puncti-, pungenti-	pricked
stoechado-, stoecho-	-farius, ordini-, serie-	rank, row
stomato-, -stoma	ori-	mouth
stonycho-	acuti-	sharp point
stortho-, storthyngo-	cuspidi-	point, spike
storyne-	cuspidi-	point, spike
strango-	suffocati-	choked
streblo-, stremmato-	torti-	twisted
strepho-, strepsi-, strepto-	torti-	twisted
strigo-	strigi-	owl
strobilo-	coni-, strobili-	pine-cone, top
strobo-	vortici-	eddy, whirl
stromato-	lamelli-, strati-	bed, bedding, layer
strombo-	turbini-	top
strombo-	cochlei-, conchi-	snail
stromne-	straguli-, tegmini-, vesti-	covering, quilt
stromne-	cubili-, lecti-, lectuli-	bed, couch, mattress
strongylo-	orbiculati-	round
stropho-	torti-	twisted
strutho-	struthioni-	ostrich
stryphno-	astringenti-	constricting
stygno-	tristitii-	gloom
stylo-	columni-	pillar, pole
styphelo-, stypho-	astringenti-	constricting
stypo-	stipiti-, stirpi-	stem, stump, block
stypsi-, styptico-	astringenti-	constricting
syco-	fici-	fig
symmetro-	proportioni-	in proportion to
symphyo-	coaliti-, connati-	grown together
symploco-	intexti-	interwoven
syll-, sym-, syn-	co-, simul-	together

31

syrrh-, sys-, syz-	co-, simul-	together
synapsi-, synapto-	iuncti-	joined together
synopsi-	conspectu-	seen together
syringo-, -syrinx	tubi-	pipe
syrrheo-	confluxi-	flowing together
systemato-	ordinationi-	organization
tachy-	celeri-, citi-, rapidi-	quick, rapid, swift
taenio-	fasciari-, liguli-, lori-	band, strap
talaeporo-	miserabili-	wretched
tany-	extenti-	stretched out
tao-	pavoni-	peacock
tapeino-, tapino-	demissi-, humili-	low, poor
tapeto-	straguli-, tapeti-	carpet, rug
tapho-	funeri-, sepultura-	burial
taphro-	fossi-	ditch
tapido-	straguli-, tapeti-	carpet, rug
tarakto-	turbati-	disturbed, troubled
tarpho-	dumeti-	thicket
tarso-	planti-	sole
tauro-	bovi-	bull
tauto-	(idem)	identical
taxi-	collocati-, dispositi-	arrangement
teicho-, toicho-	muri-, parieti-	wall
teino-, tino-	tensi-, tenti-	stretched
telamono-	baltei-, cinguli-, lori-, zoni-	band, belt, strap
tele-	longe-	far
teleio-, teleo-, telo-	fini-, ultimi-	end
teleuto-	fini-, ultimi-	end
telmato-	paludi-	marsh, swamp
telmato-	lacu-, stagni-	pond, pool
temno-	secti-	cut
tenago-	vadi-	shallows, shoal
tephro-	cinerei-	ash-grey
terato-	monstrosi-	prodigy
-teres, -teretes	custodi-	guard
terpno-, terpsi-	amoeni-	agreeable, pleasant
tetano-	extenti-	stretched out
tettigo-	cicada-	cicada
teuthido-	loligini-	squid
teutho-	sepia-	cuttlefish
thalamo-	cubiculi-	bedchamber
thalasso-	mari-	sea
thallo-	rami-	branch
thamino-	frequenti-	crowded
thamno-	frutici-	shrub
thanato-, -thanasia	morti-	death
thaumasto-	mirabili-, miri-, mirifici-	marvellous, wondrous
thaumato-, -thauma	mirabili-, miri-, mirifici-	marvel, wonder
-theca	-capsa	case
thele-	papilli-	nipple
thely-	foeminei-	female

themono-	cumuli-	heap
Theo-	Dei-	God
therei-, thero-	aestati-	summer
thermo-	calidi-	hot
thero-, therio-	feri-	wild beast
thino-	arena-, sabuli-	sand
thladia-	spadoni-	eunuch
thlipso-	pressi-	press
tholo-	testudini-, tholi-	dome
thomingo-, -thominx	funiculi-, resticuli-	cord, string
thoraco-	lorica-	breastplate
thrasy-	audaci-	bold
thraulo-, thrausto-	fragili-	breakable, brittle
threpsi-, threpto-	aliti-	nourished
thrinaco-, -thrinax	tridenti-	trident
thripo-, -thrips	cossi-	woodworm
-thrix	coma-, crini-, pili-, villi-	hair
thrombo-	grumi-	clot, lump
thryallido-	fili-	wick
thrymmato-	frusti-	bit, piece
thrypsi-, thrypto-	fracti-	broken
thyello-	procella-	storm
thylaco-, -thylax	culei-, folli-, folliculi-	bag, pouch, sack
thylaco-, -thylax	marsupi-, sacci-	bag, pouch, sack
thylaco-, -thylax	sacculi-, uteri-	bag, pouch, sack
thymeli-	ara-	altar
thyo-, -thyodes	suavi-	fragrant
thyra-, -thyra	ianua-, -ianua	door
thyreo-	scuti-	oblong shield
thyrso-	vimini-, virga-	wand
thysano-	fimbri-	fringe
tigro-	tigri-	tiger
tipho-	stagni-	pond, pool
titano-	calci-	lime
titano-	creta-	chalk
tittho-	papilli-	nipple
tmesi-, tmeto-	secti-	cut
tolypo-	glomeri-	ball of wool
tomo-	secti-	cut, piece, slice
topo-	loci-	place
toxeumato-	sagitti-	arrow
toxico-	toxico-	arrow-poison
toxo-	arci-	bow
trachelo-	cervici-, colli-	neck
trachy-	hirsuti-, hirti-, villi-	shaggy
trachy-	rudi-	coarse
trachy-	asperi-, horridi-	rough, rugged
trago-	hirci-	he-goat
trapezo-	mensa-	table
trapheco-	trabi-	beam
-trema	foramini-	aperture
tremato-, treto-	foraminati-, perforati-	perforated

trepho-	cibi-, pabulo-, nutrimenti-	food, feed
trepo-	flexi-, torquei-, versi-	turn
trepsi-, trepto-	flecti-, torti-, verti-	turned
tresi-	foramini-	perforation
tribaco-, tribo-	triti-	rubbed, worn
tribolo-	tribuli-	caltrops
tricho-	coma-, crini-, pili-, villi-	hair
trichio-	puberuli-	small hair
triemi-	sesqui-	one-&-a-half
trigono-	triquetri-	3-cornered
trocho-	roti-	wheel
trocto-	rosi-	gnawed
troglo-	foramini-	hole
trogo-	curculioni-	weevil
tropalo-	fasci-	bundle
tropho-, -trophe	nutrimenti-	food
tropi-, tropido-	carina-	keel
tropo-, -trope	versi-	change, turn
troxi-	morsu-, rodenti-	gnawing
tryblio-	crateri-	bowl
trychero-, trychino-	laciniati-	ragged
trycho-, tryo-, tryso-	fricti-, triti-	rubbed, worn
trygo-	amurca-, faeci-	dregs, lees, sediment
trymalio-, trymatio-	foramini-	little hole
trypano-	terebra-	auger, gimlet
trypeto-	excavati-, perforati-	bored
trypeto-	terebrati-	bored
tryphero-	molli-	soft
trypho-	fracti-, succisi-	broken, cut off
tyle-, tyleio-	culciti-	bolster, cushion, mattress, pillow, quilt
tylo-	bulla-, glaeba-, nodi-	callus, knob, knot, lump
tymbo-	tumuli-	mound
tympano-	tympani-	drum
tynno-	exigui-, minuti-	minute, tiny
tynno-	parvi-, pusilli-	minute, tiny
typhlo-	caeci-	blind
typho-	fumi-	smoke
typo-	typi-	blow, copy, figure, image
tyro-	casei-	cheese
tyrsio-	turri-	tower
tyttho-	parvi-	small
ulo-	cuncti-, toti-	whole
urano-	caeli-, coeli-	heaven
uro-, -ura, -oura	caudi-, caudati-	tail, -tailed
xantho-	lutei-	deep yellow
xeno-	alieni-, peregrini-	foreign, stranger
xero-	aridi-, sicci-	dry
xesto-	levi-, politi-	polished, smooth
xiphe-	ensi-	straight sword

xutho-	fulvi-	tawny, yellow-brown
xylo-, -xylon	ligni-	wood
xyro-	novaculi-	razor
xysmato-	ramenti-, scobi-, stringenti-	filings, shavings
xystero-	limi-, scalpri-	file
xysto-, xystro-	rasi-	scraped
za-	per-	very
zangklo-	falci-	sickle
zephyro-	cauri-, cori-	northwest
zeugitido-	jugali-, jugati-	yoked
zeuglo-	jugi-	yoke-attachment
zeugmato-	ligamenti-	band
zeukto-	jugali-, jugati-	yoked
zeuxi-	jungenti-	yoking
zoe-	vita-	life
zomato-	cincti-	girdled
zomo-	condimenti-	sauce
zono-	cinguli-	belt, girdle
zoo-	animali-	animal
zophero-	fusci-	dusky
zopho-	tenebri-	darkness
zostero-	baltei-	baldric, belt
zosto-	cincti-	girdled
zygo-	jugi-, -jugus	yoke
zymo-	fermenti-	leaven, yeast

35

LATIN	ENGLISH	GREEK
(iam, nunc)	now	arti-
(idem)	same	homo-
(idem)	identical	tauto-
(insula parva)	islet	nesido-
(no Latin equivalent)	flower-cup	calyco-
(no Latin equivalent)	cloak	chlamydo-
(no Latin equivalent)	water-repellent	adianto-
(no Latin equivalent)	Y-shaped	hypsilo-
(no Latin equivalent)	moustache	mystaco-
(no Latin equivalent)	S-shaped	sigmato-, sigmoideo-
(quisque)	each, every	ecasto-
(sine)	without	lipo-
ab-	away from	apo-
abditi-, arcani-, conditi-, occulti-, remoti-, secreti-, tecti-	secret	lathro-
abditi-, arcani-, occulti-, remoti-, secreti-	secret	crypsi-, crypso-
abditi-, celati-, occulti-	hidden	ceutho-
abditi-, celati-, occulti-	hidden	crypto-
accipitri-	hawk	hieraco-
-aceous, -ago, -eus	likeness	-ites, -itis
acetabuli-	cup	cyatho-
acinaci-	sabre	machairo-
acini-	grape	rhago-
acini-	grape	staphylo-
acri-	keen, pungent, sharp	drimy-
acri-	sharp	oxy-
acu-	needle	-belone
acu-	pin	perono-, -perone
acu-	needle	rhaphido-
aculei-	point	-glochin
aculeoli-	point (dim.)	glochidio-
acumini-	point	centro-, -centron
acumini-, apici-, cacumini-, summi-	point, summit, tip	-acme
acumini-, cuspidi-, mucroni-, puncti-	point	aci-, acido-
acumini-, cuspidi-, mucroni-, puncti-	point	aechmo-, aichmo-
acuti-	pointed	cestroto-
acuti-	acute	oxy-
acuti-	sharp point	stonycho-
adipi-, pingui-	fat, grease	piaro-
adipi-, pingui-	fat, grease	pimeleo-, pimelo-
adminiculi-, columini-, firmamenti-, fulcimenti-, pedamenti-	prop, stay, support	ereismato-
adunci-	hooked	grypo-
adversi-, contra-	opposite	enantio-

aedi-	house	-oecium(-oikion)
aegri-, morbidi-	diseased	astheno-
aequi-	equal	iso-
aequi-, pari-	even	artio-
aeri-	bronze, copper	chalco-
aeri-	air	meteoro-
aestati-	summer	therei-, thero-
aeterni-, immortali, sempiterni-	everlasting, undying	athanasio-, athanato-
aggeri-, tumuli-	bank, mound, hillock	ochtho-
aggregati-, coacervati-, coarctati-, conferti-, congesti-, conglomerati-, crebri-	close, crowded	adino-
agni-	lamb	amno-, arno-
agresti-	field, country	agro-
ala-	wing	pterygo-, -pteryx
alati-	winged	ptero-
alauda-	lark	corydo-
albi-	white	argi-, argo-
albi-	white	leuco-
algi-	seaweed	phyco-
alieni-, peregrini-	foreign, stranger	xeno-
alii-	other	allo-
aliti-	nourished	threpsi-, threpto-
allii-	garlic	scorodo-
alterni-	exchange	enallago-
alti-, altitudini-, ponti-, profundi-	deep (the), depth	bysso-, bytho-
alti-, elati-, proceri-	high	hypselo-, hypsi-
alti-, profundi-	deep	bathy-
altiori-	higher	hypsitero-
altissimi-	highest	hypsisto-
aluta-, cori-, pelli-	leather	byrso-
aluta-, cori-, pelli-	leather	scyto-
aluta-, cori-, pelli-	hide, leather, skin	spato-
aluti-, corii-, pelli-	leather	diphthero-
alvei-	trough	leno-
alvei-	hollowed	scapho-
alvei-, cavi-, lacuna-	hollow	cystho-
alveoli-, capsula-	box, container	cypseli-
amabili-, amoeni-, venusti-	lovely	erato-
amanti-	loving	philo-, -philus
amari-	bitter	ateramno-
amari-	bitter	picro-
ambagi-, labyrinthi-	maze	labyrintho-
ambi-	on both sides	amphi-
amni-, flumini-, fluvii-	river	potamo-
amoeni-	agreeable, pleasant	terpno-, terpsi-
amoeni-, grati-, iucundi-	agreeable, pleasant	hedy-
amplecti-	clinging	aspazo-
amplexi-	embrace	aspasio-

ampli-	ample	perisso-
ampulla-, lagena, laguncula-	flagon, flask	lageno-
ampulla-, lagena, laguncula-	flagon, flask	lagyno-
ampulli-	flask	lecytho-
amurca-, faeci-	dregs, lees, sediment	trygo-
amyli-	starch	amylo-
anati-	duck	nesso-, netto-
ancipiti-, duplici-	double	diplo-
ancora-	anchor	ankyro-
anfractuosi-	crooked	ancyclo-
angui-, serpenti-	snake	herpeto-
anguilla-	eel	enchelyo-
angulati-	angled	gonio-
anguli-	angle	-gonia
angusti-	narrow	lagaro-
angusti-	narrow	steno-
anima-	spirit	pneumato-
animali-	animal	zoo-
animi-	breath, life, soul	psycho-
annotini-	last year's	perysino-
annui-	yearly	epeteio-, eteio-
annuli-	ring	cyclo-
annuli-, circuli-	ring	crico-
anseri-	goose	cheno-
ansi-, capuli-, manubri-	handle	steleo-
antri-, cavi-, caverni-, specu-, spelunci-	cave	antro-
api-	bee	melisso-, melitto-
apici-	top	colopho-
apri-	wild boar	capro-
aqua-	water	hydato-, hydro-
aquatici-, aquatili-	aquatic	enydro-, enygro-
aquili-	eagle	aeto-
ara-	altar	thymeli-
aranei-	spider	arachno-
aratri-, vomeri-	plough	arotro-
arbori-	tree	dendro-
arca-	box	larnaco-
arca-, capsa-, cista-, pyxi-	box	ciboto-
arci-	bow	toxo-
arcuati-	looped	brochido-
ardei-	heron	erodio-
arena-, sabuli-	sand	psammo-
arena-, sabuli-	sand	thino-
areni-	sand	ammo-
argenti-	silver	argyro-
argilla-	clay	argillo-
argilla-, creta-	clay	ceramo-
aridi-, sicci-	dry	azaleo-
aridi-, sicci-	dry	xero-
arieti-	ram	crio-

armati-	armed	enoplo-
aromati-	spice	aromato-
arti-, articuli-	joint	arthro-
articulati-	jointed	enarthro-
articuli-	knuckle	condylo-
arto-, membri-	limb	gyio-
artu-, membri-	limb	colo-
arvi-	field	arouro-
ascendenti-, acclivi-	ascending	anophero-
asili, tabani-	gadfly	oistro-
asini-	ass	ono-
aspectu-	glance, look	blemmato-, -blemma
asperi-	harsh	ateramno-
asperi-	rough	cerchno-
asperi-, horridi-	rough, rugged	trachy-
assuli-, scinduli-	splinter	schidio-
assurgenti-	rising	anabaeno-
astaci-	lobster	astaco-
astaphi-	raisin	staphido-
astri-, stelli-	star	phostero-
astringenti-	constricting	stryphno-
astringenti-	constricting	styphelo-, stypho-
astringenti-	constricting	stypsi-, styptico-
atri-	dull black	melancho-, melano-
atro-	black, dark	pelio-, pelidno-
aucti-	increasing	aexi-
auctu-, incrementi-	growth	-phye(s)
audaci-	bold	thrasy-
aura-	breeze	aura-
aurei-, aureo-	gold	chryso-
auri-	ear	oto-
auriculi-	little ear	otio-
aurora-, diluculi-	dawn	orthrio-, orthro-
aurori-	dawn	eo-
australi-, meridionali-	south	noto-
autumnali-	autumnal	metoporino-
autumnali-	autumn	phthinoporo-
aveni-	oats	aegilopi-
avi-	bird	orneo-, ornitho-
axi-	axle	axono-
axilli-	armpit	maschalo-
bacca-, -coccus	berry	coccy-, -coccos
bacci-	berry	rhago-
bacilli-, virga-	rod	rhabdo-
bacilli-, virga-	rod	rhapido-
baculi-, bacilli-	cane, club, cudgel, staff	bactero-, bactro-
baculi-, fusti-	cudgel	cordyle-
baculi-, fusti-, scipioni-	staff	sceptro-
baltei-	baldric, belt	zostero-
baltei-, cinguli-, lori-, zoni-	band, belt, strap	telamono-
barba-	beard	geneiado-

barba-	beard	pogon(o)-, -pogon
basi-	base, pedestal	bathro-
bene-	well	eu-
bili-	bile	chole-, cholo-
binati-	paired	dicha-, dicho-
bini-, gemini-	twin	didymo-
bipalii-	mattock	marro-
bombyci-	silkworm	bombyco-
boni-	good	agatho-
botuli-	sausage	allanto-
bovi	ox	boö-, bou-, bu-
bovi-	bull	tauro-
brachii-	arm	brachio-
bractei-, catilli-, lamini-, patelli-	plate, sheet	elasmo-
brevi-	short	brachy-
brevi-	short	conto-
brevi-, demissi-, exigui-, humili-, minuti-, parvi-, pusilli-	low, short, small, tiny	elachy-
buccina-	bugle, trumpet	salpingo-, -salpinx
bufoni-	toad	phryno-
bulbi-	bulb	bolbo-
bulla-, glaeba-, nodi-	callus, knob, knot, lump	tylo-
bullati-	blistered, blown out	pemphigo-
bullati-	bubbled	pemphigo-
butyri-	butter	butyro-
byssi-	fine linen, muslin	sindono-
caduci-	fall	pipto-
caduci-	fall	ptoseo-, -ptosis
caeci-	blind	typhlo-
caeli-, coeli-	heaven	urano-
caelibi-	unmarried	agamo-
caerulei-	dark blue	cyano-
calcari-	spur	centro-, -centron
calcari-	goad, spur	kentro-, -kentron
calcari-	goad, quill, spur, sting	plectro-
calce-	sandal, shoe	pedilo-
calcei-, caliga-, cothurni-	boot, shoe	crepido-
calceoli-, crepidi-, solei-	sandal, slipper	crepido-
calci-	heel	pterno-, -pterna
calci-	lime	titano-
calculi-	pebble	cachleco-
calculi-	pebble	chalico-
calculi-	pebble	psepho-
calici-	chalice	depao-, -depas
calici-, poculi-, scyphi-	cup	cylico-
calidi-	hot	thermo-
calli-	track	ichno-
calli-, semita-, tramiti-	path, track	celeutho-
calvi-, glabri-	bald	phalacro-

calvi-, glabri-	bald	psedno-
calvi-, glabri-	bare	psilo-
cameli-	camel	camelo-
cammari-, gammari-	lobster	cammaro-
campani-	bell	codono-, -codon
cancelli-	lattice, trellis	cinclido-
cancri-	crab	carcino-
cancri-	crab	grapsaio-
cani-	dog	cyno-
cani-, incani-	grey, grizzled	polio-
capilli-, crini-	hair	etheiro-
capilli-, crini-, pili-	hair	come-, como-, -come
-capitata, -ceps	-headed	-cephala, -e
capiti-	head	cephalo-
capra-	she-goat	chimaero-
capra-, capri-	goat	aego-, aegi-
capreoli-, clavicula-	tendril	helino-
-capsa	case	-theca
capsi-	box	pyxido-, -pyxis
carbasi-, veli-	sail	histio-, histo-
cardinali-	hinged	ginglymato-
cardini-	pivot	cnodaco-
cardini-	hinge	ginglymo-
cardini-	pivot	polo-
carina-	keel	tropi-, tropido-
carini-	keel	steira-
carni-	flesh, meat	broto-
carni-	flesh	crea-, creio-, creo-
carni-	flesh	sarco-
casei-	cheese	tyro-
casi-	fallen	ptoto-
cassi-	net	arcy-
cassidi-, galea-	helmet	corytho-
casti-, immaculati-, puri-	spotless	aspilo-, aspiloto-
casti-, puri-	clean, pure	catharo-
catelli-	pup	scylaco-
cateni-	chain	halysi-
catilli-, catini-	dish	lopado-
cauda-	tail	cerco-
caudi-, caudati-	tail, -tailed	uro-, -ura, -oura
caudici-	stock	pythmeno-
caudici-	stock	stelecho-
caudici-, stipiti-	stump, trunk	premno-
cauli-, -caulis	stem	caulo-, -caulon
cauri-, cori-	northwest	zephyro-
causa-	cause	aetio-
caverni-	cave	spelaio-
cavi-	hollow	cyto-
cavi-	hollow	cyttaro-
cavi-	hollow, a	gyalo-
cavi-	hollow	koilo-
celebri-	celebrated	-cles

41

celeri-, citi-, rapidi-	quick, rapid, swift	tachy-
celeri-, citi-, rapidi-, veloci-	swift	ocy-
celluli-	cell	cyto-
cera-	wax	cero-
cerebri-	brain	encephalo-
cervici-	neck	isthmo-
cervici-, colli-	neck	trachelo-
ceti-	sea-monster	ceto-
charta-	paper	charto-
charta-	paper	papyro-
chasmati-, fissuri-, hiati-, rimi-	chasm, cleft	barathro-
chorda-	catgut	chorde-, chordo-
cibi-, pabuli-	food	bromato-, -broma
cibi-, pabuli-	food	sitio-
cibi-, pabulo-, nutrimenti-	food, feed	trepho-
cicada-	cicada	tettigo-
cicatrici-	scar	oteilo-
ciconii-	stork	pelargo-
cilii-	eyelash	blephari-, blepharido-
cimici-	bedbug	corio-
cincinni-	curl	bostrycho-
cincinni-, cirri-	curl	eligmato-
cincinni-, cirri-	curl	heligma-
cincinni-, cirri-	curl	plocio-, ploco-, -ploca, -ploce
cincti-	girdled	zomato-
cincti-	girdled	zosto-
cinerei-	ash-grey	spodio-
cinerei-	ash-grey	tephro-
cineri-	ashes	spodo-
cinguli-, zoni-	belt, girdle	stelmono-
cinguli-, zoni-	belt, girdle	zono-
circulari-, rotundati-	round	gyro-
circum-	around	peri-
cista-	box, chest	cisto-
citrini-	lemon-yellow	citrino-
clari-	bright	liparo-
classi-	class	phylo-
clausi-	closed, shut	cleisto-
clavi-	key	cleido-
clavi-	club	cordyle-
clavi-	club	coryne-, coryno-, -coryne
clavi-	nail	gompho-
clavi-	club	rhopalo-
clipea-	round shield	aspido-, -aspis
clipeo-	small round shield	pelto-
clivi-, colli-	hill	colono-
cluni-	rump	glouto-
cluni-	rump	pygo-
co-, col-, com-, con-, cor-	together with	hama-

co-, simul-	together	syll-, sym-, syn-, syrrh-, sys-, syz-
coacervati-, collecti-	heaped up, stored up	keimelio-
coacervati-, congesti-, cumulati-	heaped up	corysto-
coacti-	felt	pileo-, pilo-
coaliti-, connati-	grown together	symphyo-
coccinei-	scarlet	coccino-
cochlea-	snail	cochlio-
cochleari-	spoon	mystro-
cochleati, torti-	twisted	helicto-
cochlei-	spiral	helico-
cochlei-, conchi-	snail	strombo-
coeni-, caeni-	dirt	borboro-
colli-	hill	bouno-
collocati-, dispositi-	arrangement	taxi-
collyrii-, unguenti-	ointment	myro-
-color	colour	-chroma, -chroos
colorati-	coloured	chromato-
columni-	pillar	ciono-
columni-	pillar	stelo-, -stele
columni-	pillar, pole	stylo-
coma-, crini-, pili-, villi-	hair	-thrix
coma-, crini-, pili-, villi-	hair	tricho-
comini-, propinqui-	near, nigh	schedo-
commissuri-	joint	armo-
communi-, vulgari-	common	coeno-
compacti-	close-pressed	stipho-, stiphro-, stipto-
concha-	mollusc, shell	concho-
condimenti-	sauce	zomo-
conferti-	close-packed	araro-
conferti-	crowded	athro-
confluxi-	flowing together	syrrheo-
confusi-	confused, disordered	asystasio-, asystato-
coni-	cone	cono-
coni-, strobili-	pine-cone, top	strobilo-
coniuncti-	junction, union	arma-, arme-
connati-	united	gamo-
conspectu-	seen together	synopsi-
constrictioni-	constriction	sphinxi-
contempti-, despicati-, humili-, spreti-	despised	atimeto-
continenti-	mainland	epeiro-
contusi-	beaten, bruised	copto-
contusi-	pounded, struck	copto-
copiosi-	abundant, copious	perisso-
corbi-	basket	calatho-, -calathus
corbi-	basket	larco-
cordi-	heart	cardio-
corii-, pelli-	hide	phorino-
cornei-	horny reptile-scale	pholido-, -pholis
corni-, cornu-	horn	kero-

43

cornici-	crow	corone-
-cornu	horn	-ceras
cornu-, cornui-	horn	cero-
cornuti-	horned	cerato-
corona-	crown	-stemma
corona-	crown	stephano-, -stephus
corpori-	body	somato-
corpulenti-, crassi-, obesi-	bulky, stout, thick	hadro-
cortici-	bark	phloeo-, phloio-
cortici-, crusta-, folliculi-, gluma-, putamini-, testa-	husk, rind, shell	celypho-
corvi-	raven	coraco-
cossi-	woodworm	thripo-, -thrips
costa-	rib	pleuro-
coturnici-	quail	ortygo-
coxi-	hip	ischio-
crassi-, spissi-	thick	pachy-
crateri-	bowl	cratero-
crateri-	bowl	lecano-
crateri-	bowl	scaphido-
crateri-	bowl	tryblio-
cremati-, incensi-, tosti-, usti-	burnt	causto-
crepitu-	rattle	crotalo-
crepitu-, fremitu-, strepitu-	clamour, din, noise	celado-
creta-	chalk	titano-
creti-	growth	auxi-
cribelli-, cribri-	sieve	coscino-
cribri-	colander, sieve, strainer	ethmo-
crist(at)i-	crest	corytho-
crista-	crest, tuft	crobylo-
crista-	crest	lopho-
crocei-	saffron-coloured	crocoto-
cruci-	cross	stauro-
cruenti-, sanguinei-	bloody	haemato-
cruri-	leg	scelo-
crusta-	shell	ostraco-
crusta-	rind	scirrho-
cubi-	cube	cubo-
cubiculi-	bedchamber	domatio-
cubiculi-	bedchamber	thalamo-
cubili-	couch	clino-, -cline
cubili-, lecti-, lectuli-	bed, couch, mattress	stromne-
cubiti-, ulni-	forearm	pecheo-, pechy-
cubiti-	elbow	olene-, oleno-, olecrano-
cuculi-	cuckoo	coccygo-
cuculli-	cowl, hood	(no Gr. equiv.)
culciti-	mattress	stibado-
culciti-	bolster, cushion, mattress	tyle-, tyleio-
culciti-	pillow, quilt	tyle-, tyleio-
culei-	leathern sack	coryco-

culei-, folli-, folliculi-, marsupi-, sacci-, sacculi, uteri-	bag, pouch, sack	thylaco-, -thylax
culici-	gnat	conopso-
culmini-, summi-	top	acro-
culmini-, summi-	top	corymbo-
culmini-, summi-	top	corypho-
cumuli-	heap	themono-
cumuli-, tumuli-	heap, mound	soro-
cuncti-, toti-	whole	ulo-
cunei-	wedge	spheno-
cupuli-	cup	-calathus
cupuli-	cup	cotylo-, -cotyle
cupuli-	cup	scypho-
curculioni-	weevil	cio-
curculioni-	weevil	trogo-
-currens	running	-dromos
cursu-, itineri-, via-	road, way	celeutho-
curti-	curtailed, shortened, stunted	colobo-
curv(at)i-, pandi-, pravi-	bent, crooked, curved	coronido-
curvati-	curved	cyrto-
curvatura-	felloe	sotro-
cuspidi-	point, spike	stortho-, storthyngo-, storyne-
custodi-	guard	-teres, -teretes
custodii-	guard	phylaco-, -phylax
cuti-, pelli-, -pellis	skin	dermato-, dermo-, -derma
cuti-, tunica-	peel, rind	lepismato-, lepyro-
cuti-, tunica-	peel, rind	lopo-
cygni-	swan	cycno-
cylindri-, scutuli-	roller	cylindro-
cymba-	boat	cymbi-
decussati-	crosswise	chiasmato-
deflorescenti-, marcescenti-	fading, withering	maranto-
Dei-	God	Theo-
delineati-	drawn	grapho-, grapto-
delphini-	dolphin	delphino-
demissi-, humili-	low-growing, on the ground	chamae-, chamelo-
demissi-, humili-	low, poor	tapeino-, tapino-
densi-	dense	pycno-
denti-	tooth	odonto-
deserti-	desert	eremo-
dextra-	righthand	dexio-
differenti-	different	diaphoro-
digiti-	finger	dactylo-
digni-	worth	axi-, axio-
diluculi-	gloaming	lycophoto-
diri-, terribili-	dreadful	deino-
dissimili-	different	hetero-
diurni-	day	hemero-
dodranti-	span	spithame-

dolabri-	chopper, cleaver	copido-
dolabri-	pickaxe	orygo-
dolii-	jar	pitho-
dolio-	jar	amphoreo-
donati-	gift	doro-
dorsi-, tergi-	back	noto-
draconi-	dragon	draconto-
duci-	guide	hegemono-
duci-	guide	hodego-
dulci-	sweet	glyco-
dulci-, suavi-	delightful, sweet	hedy-
dumeti-	thicket	lochmaio-
dumeti-	thicket	tarpho-
dumeti-, fruticeti-, silvuli-	coppice, copse, thicket	drymo-
duplicati-	doubled, folded	diptycho-
duplici-	double	disso-, ditto-
duri-	hard	aageo-
duri-	hard	ateramno-
duri-	hard	sclero-
e-, ex-	without	a-, an-
eduli-, esculenti-	eatable	brosimo-
efficaci-	active, vigorous	drastico-
eiecti-	throw out	ecballo-
eleganti-	neat	compso-
elevati-	raised	hyperphoro-
-ellus, -illus, -ulus	(diminutives)	-idion
enormi-	abnormal, irregular	anomalo-
ensi-	straight sword	xiphe-
equi-	horse	hippo-
equulei-, equuli-	colt, foal, pony	hippario-
erinacei-	hedgehog	chero-
erinacei-	hedgehog	echino-
erinacei-	hedgehog	schyro-
essenti-, natura-, vita-	being, essence, existence	hyparxeo-
essenti-, natura-, vita-	being, essence, existence	onto-
esu-	eating	phago-
excavati-	hollowed	glaphyro-
excavati-, -cavus	hollow	coelo-
excavati-, perforati-, terebrati-	bored	trypeto-
exigui-	insignificant	asemanto-
exigui-, minuti-, parvi-, pusilli-	little, small	baeo-
exigui-, minuti-, parvi-, pusilli-	minute, tiny	tynno-
expulsi-	driven away	elaterio-
extensi-, extenti-	stretched	ecteino-, ecteno-
extenti-	stretched out	tany-
extenti-	stretched out	tetano-
extra-	outside	ecto-, exo-
extra-, ultra-	away from, beyond	exo-

extra-, ultra-	out of, without	exo-
extremi-	outermost	pymato-
exuti-, exuvi-	slough off	ecdysi-
faci-, funali-, taeda-	torch	lampado-
faciei-	appearance	-opsis
-factus	made	-poietos
falci-	sickle	drepano-
falci-	sickle	zangklo-
falconi-	falcon, harrier, hawk	circo-
falculi-	billhook	copido-
falsi-, spurii-	false, spurious	pseudo-
farcti-	stuffed	sacto-
farini-	floury, mealy	aleuro-
-farius, ordini-, serie-	rank, row	sticho-, -stichos
-farius, ordini-, serie-	rank, row	stoechado-, stoecho-
fasci-	bundle	angkalo-
fasci-	bundle	-desme
fasci-	bundle	tropalo-
fasci-, fasciculi-	bundle, fagot	phacelo-
fasci-, manipuli-, mergiti-	sheaf, truss	-dragma
fasciari-, liguli-, lori-	band, strap	taenio-
fastigi-	gable, pinnacle, roof	pterygio-
fastigiati-	tapering	phoxi-
fauci-	throat	laimo-
fauci-	throat	laryngo-
fauci-	throat	pharyngo-
favi-	honeycomb	cerio-
feli-	cat	ailuro-
femori-	thigh	mero-
-fer, -ger	bearing, carrying	-phoros
ferenti-	bringing	phere-
feri-	wild beast	thero-, therio-
fermenti-	leaven, yeast	zymo-
ferri-	iron	sidero-
feruli-, paxilli-, virga-	peg, rod	cercido-
fibra-, nervi-	fibre, sinew	inio-
fibrati-, nervosi-	fibrous, sinewy	-inodes
fibri-	beaver	castoro-
fici-	fig	ischado-
fici-	fig	syco-
-ficiens	making	-poiea, -poiesis, -poios
fictili-, figlini-	pottery	chytro-
fili-	thread	mito-
fili-	thread	nemato-
fili-	thread	rhammato-
fili-	wick	thryallido-
filici-	fern	pteri-, pterido-
fimbria-	fringe	cosymbo-
fimbri-	fringe	lomato-, -loma
fimbri-	fringe	thysano-
fimbriati-	fringed	crossoto-

47

fimi-	dung	ontho-
fimi-	dung	scato-
fini-	boundary	horio-, horismo-
fini-, ultimi-	end	teleio-, teleo-, telo-
fini-, ultimi-	end	teleuto-
-firmamentum	prop, stalk, support	-pelma
firmi-, solidi-, stabili-	firm, lasting, stout	ochyro-
firmi-, stabili-	holding fast to	aptosi-, aptoto-
fissi-	cracked, split	rhexi-
fissi-, scissi-	split	schis-, schismato-, schisto-
fissi-, scissi-	split	schizo-
fissura-, hiatu-, rima-	chasm, cleft, gully, ravine	pharango-
-fixus, -fixa	fixed, made fast, secured	pegmato-, -pegia
flabelli-	fan	rhipi-, rhipidi-
flaccidi-	flabby	chauno-
flagelli-	whip	mastigo-
flagr-, flammi-, igni-, incensi-	fire	pyro-
flammei-, ignei-	fiery red	pyrso-
flecti-	bent	campso-, campto-, campylo-
flecti-, torti-, verti-	turned	trepsi-, trepto-
flectibrachii-	bent arm	angkale-
flexi-	bending, a	campe-
flexi-	bent, crooked	scambo-
flexi-	bent	scolio-
flexi-, flecti-	bent	rhoiko-
flexi-, obstipi-	bent, crooked	rhaibo-
flexi-, torquei-, versi-	turn	trepo-
flexu-	bend	ankono-
floccosi-, lanati-	woollen	erio-
-flora	-flowered	-antha, -e
florei-, floridi-	flowery	antheio-
flori-	flower	antho-, -anthemon
fluctu-	billow, swell, wave	cymo-
flumini-	stream	rheithro-
fluxi-, flucti-	flow	rheo-
fluxi-, flucti-	flow	rheumato-
fluxi-, flucti-	flow	-rrhoe, -rrhoea, -rrhoa
fodi-	burrow, dig	orycho-
foeminei-	female	gyno-
foeminei-	female	thely-
foeni-	hay	chorto-, -chortos
foeti-, inchoati-	unborn, unformed	embryo-
foetidi-	stinking	bromo-
foetidi-, graveolenti-	stinking	ozo-
-folia	-leaved	-phylla
folii-	leaf	phyllo-
folli-	bellows	physo-
fonti-, scaturigini-	spring, a	pego-
foraminati-, perforati-	perforated	tremato-, treto-
foramini-	aperture	-trema

foramini-	perforation	tresi-
foramini-	hole	troglo-
foramini-	little hole	trymalio-, trymatio-
foramini-, fovea-, putei-	hole, pit	lacco-
forfici-	shears	psalido-
-foris, -porta	gate	pylo-, -pyle
forma-, specie-	figure, form, shape	eido-
formi-, -formis	shape	morpho-, -morpha
formica-	ant	myrmeco-
formidabili-	fear-generating	phobero-
fornaci-	kiln	camino-
fornici-	vault	camaro-
fossi-	burrowed, dug	orycto-
fossi-	ditch	taphro-
fossili-	dug	scapto-
fovei-	pit	bothrio-
fovei-, putei-	pit	barathro-
fracti-	broken	clasto-
fracti-	broken	klasto-
fracti-	broken	thrypsi-, thrypto-
fracti-, succisi-	broken, cut off	trypho-
fragili-	brittle	acampo-
fragili-	brittle	clao-
fragili-	breakable, brittle	thraulo-, thrausto-
fragili-, friabili-	brittle, crumbly, loose	psathyro-
fragranti-	fragrant	euod-, euosm-
fragranti-	pleasant-scented	hedyosmo-, hedypnoi-
fragranti-	fragrant	myristico-
fratri-	brother	adelpho-
fraudi-	cheat, impostor	phenaco-, -phenax
fremitu-	crackling, noisy, roaring	bromeso-
fremitu-, mugitu-, strepitu-	bellowing, roaring	bryche-, brycheto-
frequenti-	crowded	thamino-
fricti-, triti-	rubbed, worn	trycho-, tryo-, tryso-
frigidi-	cold	psychro-
-fringilla	-finch	-spiza
fronti-	forehead	metopo-
-fructa	fruited	-carpa
frugi-	fruit	carpo-
frumenti-	corn	sito-
frusti-	fragment, piece	clasmato-
frusti-	bit, piece	thrymmato-
frustuli-, mica-	crumb, morsel	psacado-
frustuli-, mica-	crumb, morsel	psicho-
frustuli-, mica-	crumb, morsel	psom(i)o-
frutici-	shrub	thamno-
fuci-	drone	cepheno-
fugitivi-	runaway, a	drapeto-
fulguri-	lightning	astrapo-
fulici-	coot	phalaro-
fuligini-	soot	lignyo-
fulmini-	thunderbolt	cerauno-

fulti-	prop, support	sterigmo-
fulvi-	tawny	cirrho-, cirro-
fulvi-	tawny, yellow-brown	xutho-
fumi-	smoke	capno-
fumi-	smoke	typho-
fundi-	base, bottom	pyndaco-
fundi-	base, bottom	pythmeno-
funeri-, sepultura-	burial	tapho-
fungi-	fungus	myceto-, myco-, -myces
fungi-	sponge	sompho-
fungi-	sponge	spongio-, spongo-
funi-	cable	sparto-
funi-, resti-	rope	schoeno-
funi-, resti-	rope	-sira, -seira
funiculi-, resticuli-	cord, string	thomingo-, -thominx
furcati-	forked	dicraeo-
furfuri-	bran	pityro-
furni-	oven	ipno-
fusci-	dusky	orphno-
fusci-	dusky	percno-
fusci-	dark, dusky	phaeo-
fusci-	dusky	zophero-
fusi-	spindle	atracto-
fusi-	spindle	clostero-
fusti-	cudgel	scytalo-
galli-	cock	alectoro-, alectryo-
galli-	gall	cecidio-
gelu-	frost	cryo-, crymo-
gelu-	frost	rhigo-, rhigio-
gelu-, geli-	frost	pageto-
geni-	race	phylo-
geniculati-	-kneed	gony-
genitali-	reproductive organs	-gone
genu-	knee	gony-
-genus	kind, of a	-geneus
gibberi-, gibbi-	hump, protuberance	hybo-
gibbi-	humped, hunched	cypho-
glaciei-	ice	cryo-, crymo-
glaciei-	ice	crystallo-
gladii-	broadsword	rhomphaio-
glandi-	acorn	balano-
glanduli-	gland	adeno-
glarea-	gravel, shingle	cherado-
glauci-	blue-grey	glauco-
globi-	ball	sphaero-
globosi-, glomerosi-, orbiculati-, rotundi-	round	gongylo-
glomeri-	ball of wool	tolypo-
gloriosi-	glory	-doxa
gluma-	husk	lemmato-
glutini-	glue	colla-, collo-

glutinosi-	glue	gloeo-
gracili-	graceful	charito-
gracili-	slender	ischno-
gradi-, scali-	step, stair	bathmo-
gralli-	stilt	colobathro-
gramini-	grass	-chloe
grandi-, grossi-	large	megalo-
grandini-	hail	chalazo-
gravi-	heavy	bary-
gravi-, ponderosi-	heavy	embritho-
grui-	crane	gerano-
grumi-	clot, lump	thrombo-
grumuli-	ant's-nest	myrmedono-
grumuli-	heap, hummock	soro-
gryphi-	griffin	grypo-
gubernaculi-	rudder	pedalio-
gurgiti-, profundi-, voragini-	abyss, gulf	barathro-
gutta-	exudate, ooze	stacto-
gutta-	drop	stagono-
gutta-, macula-	spot	balio-
guttati-, maculati-, punctati-	spotted	sticto-, -sticta
guttati-, maculati-, punctati-	spotted	stizo-
guttati-, maculosi-	dappled	psaro-
habitu-	figure, form, shape	schemato-
haedi-, hoedi-	kid	eripho-
hami-, unci-	hook	harpago-
hamuli-	fish-hook	ancistro-
hasti-	spear	aechmo-, aichmo-
hasti-	spear	akonto-
hasti-	spear	dory-
hebdoma-	week	hebdoma-
hianti-	gaping	chaeno-
hianti-	gaping	chasco-
hiatu-	chasm	chasmato-
hiatu-	chasm	sarmato-
hiemi-	winter	cheimo-
hinnulei-	fawn	elapho-
hinnulei-	fawn	kemado-
hinnulei-	fawn	nebro-
hirci-	he-goat	trago-
hircipelli-	goatskin	sisyro-
hirsuti-, hirti-, hispidi-, villi-	shaggy	dasy-
hirsuti-, hirti-, villi-	shaggy	trachy-
hirudini-	leech	bdello-, -bdella
hirundini-	swallow	chelidono-
homini-, viri-	man	andro-, anthropo-
homini-, viri-	man	broto-
hora-	hour	horo-
hornotini-, tempestivi-	seasonable	horaio-
horridi-	bristling	phrisso-, phrixo-
horti-	garden	cepo-

51

humeri-	shoulder	omo-
humi-, soli-	ground	edapho-
humidi-, made-	moist	hygro-
humidi-, made-	damp, moist, wet	notero-
humidi-, rori-	moisture	icmado-
humili-	low	chthamalo-
hystrici-	porcupine	hystricho-
iacti-	thrown	rhipto-
iactu-	throw	bolo-, -bolos
iactu-	throw	rhipso-
iaculi-, teli-	dart	belo-
ianua-, -ianua	door	thyra-, -thyra
icteri-	jaundice	ictero-
iecori-	liver	hepato-
igni-, flamma-	flame	phlogi-, -phlox
ignobili-	undignified	asemno-
ignoti-	unknown	adelo-
ignoti-	unknown	agnoto-, agnosto-
imagini-	likeness	eikono-
imbri-	rain, shower	ombro-
immani-	prodigious	perisso-
immani-, ingenti-, monstrosi-	huge, monstrous	peloro-
immani-, ingenti-, monstrosi-	prodigious, terrible	peloro-
immaturi-	unripe	omphaco-
immobili-	immoveable	akineto-
impari-	odd	perisso-
inaequi-	unequal	aniso-
inaequi-	uneven	perisso-
inaequi-	unequal	scaleno-
inaequi-, iniqui-	uneven	anomalo-
incolenti-	dwelling in	endemo-
incudi-	anvil	acmo-
inculti-	untilled	atropo-
incunabuli-	swathed	spargano-
indici-	pointer	gnomono-
indurati-	hardened	atyloto-
inermi-	unarmed	anoplo-
inflati-	blown out	physeto-
inflati-	blown out	pneumatico-
inflicti-	strike	plectro-
infra-, sub-	below, down	cata-
infundibuli-	funnel	choano-, chono-
innumeri-	countless	myrio-
insecti-	insect	entomo-
insigni-, praecipui-	remarkable	perisso-
insiti-, surculi-	graft, scion, slip	clemato-, -clema
insueti-	unusual	aethio-, -eo-
insuli-	island	neso-
integri-	entire, whole	holo-
integumenti-, involucri-, operculi-	covering, a	-calymma

inter-	among	meta-
interiori-	inner	esotero-
intestini-	gut	entero-
intexti-	interwoven	symploco-
intexti-, plicato-	plaited, woven	pleco-, plecto-
intimi-	innermost	esotato-
intra-	within	endo-, ento-
invisi-	unseen	aphano-
involucri-	covering	eilemato-
involuti-	rolled up	aneilemato-
ipsi-	self	auto-
irpici-	harrow	bolokopo-
iuncti-	joined together	synapsi-, synapto-
iuxta-	next to	meta-
iuxta-	beside	para-
iuxta-	near, close by	pros-
jaculi-, teli-	dart	akonto-
jugali-, jugati-	yoked	zeugitido-
jugali-, jugati-	yoked	zeukto-
jugi-	ridge	lopho-
jugi-	yoke-attachment	zeuglo-
jugi-, -jugus	yoke	zygo-
jungenti-	yoking	zeuxi-
-labiatus	-lipped	-chilus
labii-	lip	cheilo-
lacerati-	torn, mangled, rent	amyxi-
lacerati-	torn	drypto-
lacerati-	ragged, torn	rhacoi-, -rhacodes
lacerati-	torn	sparasso-, sparatto-
lacerti-	lizard	sauro-
laciniati-	ragged	trychero-, trychino-
lacrimi-	tear	dakryo-
lacti-	milk	galacto-
lacu-, stagni-	pond, pool	telmato-
laesioni-, vulneri-	injury	blabe-
laeti-	delighting in, rejoicing in	chaero-
laevi-, laevigati-	smooth	leio-
laevi-, laevigati-	smooth	lispo-, lisso-
lamelli-, strati-	bed, bedding, layer	stromato-
lamina-	blade	spatho-
lampetro-, muraeni-	lamprey	bdello-, -bdella
lancea-	lance, spear	loncho-
lani-	wool	eiro-
lani-, velleri-	wool	mallo-
lanugini-	removeable covering	-achne
lanugini-	down	chnoo-
lanugini-	down	lachno-
lapidi-, saxi-	stone	litho-
lapsu-	fall, slip	sphalmato-
laquei-	noose	brocho-

lateri-	side	pleuro-
lateri-	brick	plintho-
lati-	broad, wide	eury-
lati-	broad, wide	platy-
laxi-	loose	chalaro-
laxi-	slack	laparo-
laxi-, remissi-	slack	aneto-, aneuro-
legumini-	pulse	osprio-
leni-, miti-, molli-	soft	malaco-
leni-, miti-, molli-	mild, soft	prau-
lenti-	lentil	phaco-, phako-
leoni-	lion	leonto-
lepadi-	limpet	lepado-
lepori-	hare	lago-
lepori-	hare	ptaco-, ptoco-
levi-	lightweight	elaphro-
levi-, politi-	polished, smooth	xesto-
liberi-	free	eleuthero-
lieni-	spleen	spleno-
ligamenti-	band	sphinctero-
ligamenti-	band	zeugmato-
ligati-, stricti-, vincti-	bound	sphigmato-
ligati-, stricti-, vincti-	bound	sphincto-
ligni-	stick, wood	phrygano-
ligni-	wood	xylo-, -xylon
liguli-, lori-	band	desmo-, desmos
lili-	lily	lirio-, leirio-
limi-	mud	ily-, ilyo-
limi-, scalpri-	file	xystero-
limini-	threshold	bathmo-
linei-	line	-gramme
lingui-	tongue	glosso-, glotto-
linguli-	small tongue	glossario-, glossidio-
lini-, lintei-	flax, linen	othono-, othonio-
lintri-, naviculi-	boat	scapho-
litteri-, scripti-	letter, writing	grammato-, -gramma
lividi-	leaden	pelio-, pelidno-
lobi-	lobe	lobo-
loci-	place	topo-
locusta-	grasshopper	acrido-
loligini-	squid	teuthido-
longe-	far	tele-
longi-	long	dolicho-
longi-	long	macro-
lori-	strap, thong	himanto-
lorica-	breastplate	thoraco-
luci-	light	auge-
luci-	light	phengo-
luci-	light	photo-
ludibri-	plaything	-paegma
lumbrici-	earthworm	scoleco-
lumini-	lamp, light	lychni-, lychno-

luna-	moon	meni-
luna-	moon	seleni-
lunati-	crescentic	menoidi-
lupi-	wolf	lyco-
lutei-	quince-yellow	melino-
lutei-	deep yellow	xantho-
luti-	clay	pelo-
lynci-	lynx	lynco-
lyra-	harp, lute, lyre	cithara-
lyra-	lyre	lyro-
macula-	stain	celido-
macula-	spot, stain	spilo-
maculosi-	variegated	eusticto-
magni-	great	mega-
magnifici-	splendour	aglaio-, aglao-
maiori-	greater	meizo-
male-	bad	caco-
male-	bad, ill	dys-
male-, mali-	bad, evil	phaulo-
mali-	apple	melo-
mallei-	hammer	sphyra-
mammi-	breast	masto-
manci-	maimed	pero-
manifesti-	evident	delo-
manifesti-	evident	phanero-
manipuli-	armful	angkalido-
manu-	hand	cheiro-, chero-, chiro-
margarita-	pearl	margarito-
margini-	border	craspedo-
margini-	edge	ityo-
mari-	sea	ponto-
mari-	sea	thalasso-
mari-, masculi-	male	arrheno-, arseno-
marini-	sea, of the	enalio-
marini-	sea, of the	pelago-, pelagio-
maritimi-	seaside, by the	paralia-
marmori-	marble	marmaro-
marsupi-, sacco-	pouch	pera-
matri-	mother	metro-
maturi-	ripe	horio-
maturi-	ripe	pepeiro-
matutini-	morning	proio-
maxilli-	jaw	gnatho-
maxilli-	jawbone	siagono-
maximi-	very big	megisto-
medi-	middle	meso-
medic-	doctor	iatro-
medulla-	marrow	myelo-
meleagridi-	guineafowl	meleagrido-
melli-	honey	meli-
membrana-	membrane	-chorion

55

membrani-	thin-skinned	hymeno-
membrani-	membrane	meningo-
mensa-	table	trapezo-
mensura-	measure	metre-, metrio-, metro-
menti-	chin	geneio-
meridiei-	noon	mesembria-
mersi-	dipped	bapti-
mersi-	diving	colymbi-
minii-	vermilion	milto-
minimi-	smallest	elachisto-
minimi-	least	microtato-
minori-	less	elasso-, elatto-
minori-	less, smaller	meio-
minori-	lesser	microtero-
mirabili-, miri-, mirifici-	marvellous, wondrous	thaumasto-
mirabili-, miri-, mirifici-	marvel, wonder	thaumato-, -thauma
miserabili-	wretched	talaeporo-
miti-	mellow	pepono-
mitri-	headdress	mitro-
mitylli-	mussel	myako-
mixti-	mixed	mixo-
mola-	mill	mylo-
molli-	soft	hapalo-
molli-	soft	tryphero-
monili-	necklace	hormo-
monstrosi-	prodigy	terato-
monti-	mountain	oreo-, ores-, oro-
morsu-	bite	dako-, -dakos
morsu-, rodenti-	gnawing	troxi-
morti-	dead	necro-
morti-	death	thanato-, -thanasia
muci-	slime	borboro-
muci-	slime	myxo-
mulcti-, mulgi-, mulsi-	milking	bdallo-
multi-	much	pletho-
multi-	many	poly-
mundi-	world, universe	cosmo-
muri-	mouse	myo-, -mys
muri-, parieti-	wall	teicho-, toicho-
murici-	purple	calche-
murra-	myrrh	smyrno-
musc-	moss	bryo-
musci-	fly	myio-, -myia
musculi-	muscle	myo-, -mys
musteli-	weasel	gale-, gali-
mutati-	changed	meta-
mutationi-	change	allago-
mutui-	mutual	allelo-
nasi-	nose	myctero-
nasi-	nose	rhino-
natanti-	swimming	colymbi-

natanti-	swimming	necto-
neca-, -cidus	murder	phono-
neci-, -cidus	murder	ctono-, -ctonos
nemori-	glade	nemeo-
nervi-	fibre, ligament, nerve	neuro-
nervi-	sinew, tendon	neuro-
nigri-	black	celaeno-, celaino-
nigri-	shiny black	melancho-, melano-
nimii-	excessive	perisso-
nitenti-	shining	phaedro-, phaeno-
nitenti-	shining	phano-
nitenti-, splendenti-	glossy, polished, shining	stilbo-, stilpno-, stilpsi-
nitidi-	bright	gano-
nitidi-	bright	lampro-
nivi-	snow	chiono-
nivi-	snow	niphado-
nocenti-, nocivi-, noxii-	baneful, damaging, harmful	blabero-, blabo-
nocenti-, nocivi-, noxii-	hurtful, injurious	blabero-, blabo-
nocenti-, nocivi-, noxii-	baneful, damaging, harmful	blapto-, blaptiko-, -blapton
nocenti-, nocivi-, noxii-	hurtful, injurious	blapto-, blaptiko-, -blapton
nocti-	night	nycti-, nycto-
nodosi-, torulosi-	knobbly	gongylo-
nomini-	name	onomato-
-nota	mark	-stigma
nota-	mark	charagmato-
notati-	marked	characto-
notati-	marked	stigmato-
nothi-	bastard	notho-
novaculi-	razor	xyro-
novi-	new	caeno-, ceno-
novi-, novo-	new	neo-
nubi-	cloud	nepheo-, nepho-, nephelio-
nuci-	nut	caryo-, -caryon
nuntii-	messenger	angelo-
nuntii-	herald	keryko-
nutanti-, nutati-	nodding	brizo-
nutrimenti-	food	tropho-, -trophe
obici-, pessuli-, repaguli-, sera-	bolt	cleithro-
obici-, pessuli-, repaguli-, sera-	bolt	gompho-
obliqui-	slanting	loxo-
obliqui-	slanting	plagio-
obscuri-	indistinct	amydro-
obturaculi-, obturamenti-	bung, plug, stopper	bysmo-, bystro-
obturatori-	bung, plug, stopper	bysmo-, bystro-
obtusi-	blunt	ambly-
obtusi-	blunt	copho-
occidenti-	west	hespero-
ocelli-	eyelet	ommato-, -omma
ochracei-, sili-	ochreous, pale yellow	ochro-

ocrei-	legging	cnemido-, -cnemis
oculi-	eye	blemmato-, -blemma
oculi-	eye	ophthalmo-
odori-	perfume	myro-
odori-, olor-	smell	osm-, osmo-, -osme
olei-	oil	elaio-
olei-	oil	lipa-
olla-	pot	chytro-
olla-	jar	stamno-
omni-	all	pan-, panto-
operculi-	cover	elytro-
operculi-	cover	pomato-, pomatio-
optimi-	best	aristo-
orbiculati-	round	strongylo-
ordinationi-	organization	systemato-
ordini-	order	phylo-
ori-	opening	-poros
ori-	mouth	stomato-, -stoma
orienti-	east	anatoli-
ossi-	bone	osteo-, osto-
ovi-	sheep	melo-
ovo-	egg	oo-
pabuli-	food	sitio-
pala-	shovel	scapano-, -scapane
palea-	chaff	achyro-
pali-	stake	scolo-, scolopo-
pali-	stake	stalico-, stalido-
palliati-	cloak	-chlaena
pallii-	mantle	himatio-
palpebri-	eyelid	blepharo-, -blepharon
paludi-	marsh, swamp	telmato-
palustri-	marsh	heleo-, helo-
pani-	bread	arto-
pani-	bread	sitio-
paniculi-	reed-plume	antheli-
papilli-	nipple	thele-
papilli-	nipple	tittho-
parasiti-	sponger	colaco-
parma-	target	pelto-
parti-	part	-meros
parvi-	small	micro-
parvi-	little, small	pauro-
parvi-	small	tyttho-
pascui-	pasture	nomo-, -noma
passi-	suffering	patho-
patelli-	plate	placo-
-patina, -patella	flat dish, patella	-patane
pauci-	few	oligo-
pauci-	few	pauro-
pauci-	few	spano-
pavoni-	peacock	tao-

pectini-	comb	cteno-
pectori-	breast	sterno-
pectori-	chest	stetho-
pedamini-, pedamenti-	vine-prop	camaco-
pedi-, -pes	foot, -footed	podo-, -poda
pediculi-	louse	phtheiro-
pendenti-, penduli-	hanging	cremasto-
penicilli-	brush	callyntro-
pennati-, plumati-, plumosi-	feathered	pteno-
pennati-, plumati-, plumosi-	feathered	pteridio-, pterino-
penni-, plumi-	feather	ptilo-
per-	through	dia-
per-	very	za-
percolati-	filtered, strained	hylisto-
perdici-	partridge	perdico-
perforati-	perforated	ethmoideo-
permutanti-	exchanging	amoibo-
personati-	face	-ops
-pes	foot	-pus, -podos
petasi-	hat	petaso-
pharetra-	quiver	goryto-
physali-	wintercherry	halicacabo-
pici-	pitch	pisso-
pili-	javelin	akonto-
pisci-	fish	ichthyo-
pituita-	mucus	phlegmato-
plani-	even, level	homalo-
plani-	flat	petalo-
plani-	flat	placo-
planta-, planti-	sole of the foot	pelmato-
planti-	plant	phyto-
planti-	sole	tarso-
pleni-	full	pleo-, plero-
plica-	fold	ptysso-
plicanti-	folding, a	ptyxo-
plicati-	folded	ptycho-
plicati-	folded	ptygmato-
plicati-	folded	ptykto-
plumbi-	lead	molybdo-
pluri-	more	pleio-
plurimi-	most, very many	pleisto-
pluvii-	rain	hyeto-, -hyet
poculi-	small cup	cymbi-
poculi-	winecup	phiali-, phialo-
poculi-	winecup	poterio-
poculi-, scyphi-	goblet	cypello-
politi-	finished, polished, smooth	glaphyro-
ponti-	bridge	gephyro-
porcelli-	piglet	choiridio-
porci-	pig	choero-, choiro-
porracei-	leek-green	prasino-
post-	after	meta-

posteri-	behind, back	opistho-
postremi-	last	opistato-
postremi-, ultimi-	last	prumno-, prymno-
postremi-, ultimi-	last	pymato-
prae-	before, in front of	pro-, proso-
praecoci-	early	proio-
praecociori-	earlier	protero-
praediti-	endowed	proikto-
prati-	meadow	-leimon, -limon
pravi-	bent, crooked	ankylo-
pravi-	crooked	cyllo-
prehensi-	seizing	harpazo-
pressi-	press	thlipso-
primi-	first	proto-
principi-	chief	arche-, archi-
procella-	storm	thyello-
procella-, turbini-	hurricane	lailapo-
prominenti-	raised	execho-
proni-	lying forwards	prene-
propinqui-	near	anchi-
proportioni-	in proportion to	symmetro-
protecti-, sima-	cornice, eaves	geiso-, geisso-
proximi-, vicini-	near, neighbour	plesio-
pruina-	hoar-frost, rime	pachne-
pruni-	damson, plum	coccymelo-
psittaci-	parrot	psittacin-
puberuli-	small hair	trichio-
pulchri-, venusti-	beautiful, charming, lovely	calli-, calo-
pulici-	flea	psyllo-
pulmoni-	lung	pneumono-
pulveri-	dust	coni-, conio-, conis-
pulveri-	dust	psocho-
pumili-	dwarf	nano-
puncti-, pungenti-	pricked	stizo-
punicei-	carmine, crimson	phoeniceo-, phoenico-
purgativi-	cleansing, purgative	cathartico-
purpureo-	purple	porphyreo-, porphyro-
pustuli-, tuberculi-	pimple, tubercle	chalazo-
pustuli-, vesici-	blister, pustule, vesicle	phlyctaino-
pustuli-, vesici-, vesiculi-	blister, pustule, vesicle	psydraco-
putamini-	fruitstone	pyreno-
putationi-	pruning	cladeuto-
putrescenti-	decaying	maranto-
putridi-	decay	-phthora
putridi-	rotten	sapro-
putridi-	rotten	sathro-
racemi-, uvi-	cluster	botryo-, -botrys
radici-	root	rhizo-, -rrhiza
radii-	ray	actino-
ramenti-, scobi-, stringenti-	filings, shavings	xysmato-
rami-	branch	clado-

rami-	branch	rhadico-
rami-	branch	spadico-
rami-	branch	thallo-
ramuli-	branchlets	cladio-, cladisco-
rani-	frog	batracho-
rari-	thin	araio-
rari-	uncommon	perisso-
rari-	scanty	spanisto-
rasi-	scraped	xysto-, xystro-
re-	again	ana-
re-	again, back	palin-
recessi-	nook	mycho-
recti-	straight	euthy-
recti-	straight	ithy-
recti-	straight	ortho-
regi-	king	basileo-
regioni-	place, plot, region, space	choro-
remi-	oar	cope-
remoti-	distant	macryno-
reni-	kidney	nephro-
repenti-	creeping	herpesti-
resina-, resini-	resin	rhetino-
reti-	net	dictyo-
reti-	fishing-net, -basket	gripho-
reticuli-	hairnet	cecryphalo-
reverti-	returning, returned	anacamps-, anacampt-
rigidi-	inflexible, unbending	atropo-
rima-	break, fracture, laceration	regmato-
rima-	rent, rupture, tear	regmato-
rima-	chasm, chink, cleft	regmato-
rima-	chink, crack, rent	rhagado-
robigini-	mildew	erysibo-
robusti-, validi-	strong, stout	alkimo-
robusti-, validi-	strong	cratero-, crato-
robusti-, validi-, viri-	strength	stheno-
rori-	dew	droso-
roseo-, rosi-	rose	rhodo-
rosi-	gnawed	trocto-
rostri-	beak	rhamphi-, rhampho-
rostri-	beak, snout	rhyncho-
roti-	wheel	trocho-
rubidi-	deep red	pyrrho-
rubri-	red	erythro-
rudi-	coarse	trachy-
rufi-	orange-red	pyrrho-
rugi-	wrinkle	rhyti-, rhytido-
rugosi-	shrivelled	rhikno-
rugosi-	wrinkled	rhyso-, rhysso-
rupi-, saxi-	rock	petro-
sacci-	bag	phascolo-
sacci-	bag, sack	sacco-

61

sacri-, sancti-	sacred	hiero-
sagitti-	arrow	belo-
sagitti-	arrow	toxeumato-
sali-, salso-	salt	halo-
saponi-	soap	smegmato-
sarculi-	hoe	scalido-
scabie-	scurf	leicheno-
scabri-	rough, scabby	psoro-
scaevi-	lefthand	scaio-
scala-	ladder	apobathro-
scala-	ladder	climaco-, -climax
scandenti-	climbing	periallocaulo-
scapi-	shaft, stem	scapo-
scapuli-	shoulder-blade	omoplato-
scarabaei-	beetle	cantharo-
scarabaei-	stag-beetle	carabo-
sciuri-	squirrel	sciuro-
scopae-	besom, broom	corema-
scopae-	besom, broom	corethro-
scopi-	broom	saro-
scopuli-	cliff	cremno-
scopuli-	rock	scopelo-
scorpii-	scorpion	scorpio-
scrofa-	old sow	gromphado-
sculpti-	carved	glypho-, glypto-
scuti-	oblong shield	hoplo-
scuti-	oblong shield	thyreo-
sebi-	suet, tallow	stearo-, steato-
secti-	cut	temno-
secti-	cut	tmesi-, tmeto-
secti-	cut, piece, slice	tomo-
secundi-	second	deutero-
securi-	axe	axino-
securi-	axe	peleceo-, pelecy-, pelyco-
sedi-, sedili-	seat	-hedra
sella-	saddle	sagmato-
semi-	half	hemi-, hemisy-
-semina	-seeded	-sperma
semini-	seed	spermato-
semini-	seed	sporo-
semper-	always	aei-, ai-
semper-	ever-	empedo-
sempervirenti-	evergreen	empedophyllo-
seni-, veteri-	old	geraio-, geronto-
senti-	thorn	akaino-
sepia-	cuttlefish	teutho-
septentrionali-	north	boreo-
septi-	fence	erymno-
septi-	fence, hedge, partition	phragmo-, phragmato-, -phragma
sera-	bar	rhabdo-
seri-, sero-	fluid, juice	ichoro-

sericei-	silk	serico-
serotini-	late	hystero-
serotini-	late	opse-, opsi-
serpenti-	snake	ophio-
serrati-	jagged	carcharo-
serrati-	cut	prismato-, pristo-
serri-	saw	priono-
serti-	crown, garland, wreath	-stelma, -stemma
serti-	garland, wreath	stephano-, -stephus
sesqui-	one-&-a-half	triemi-
sessili-	seated	hedraio-
seti-	bristle	chaeto-, -chaete
sigillati-, signati-	sealed, stamped	sphragisto-
sigilli-, signi-	seal, stamp	sphragido-
signi-	mark	-sema
siliqua-	pod	celypho-
simi-	snub-nosed	simo-
simia-, simii-	monkey	cercopo-
simii-	monkey	pitheco-
simili-	similar	homoio-
simili-	like	-oides
simplici-	plain, simple	aphelo-
simplici-	simple	lito-
singuli-	once	hapax-
sinistri-	left	aristero-
sinistri-	lefthand	euonymo-
sinu-	bosom	colpo-
siri-	pit	siro-
situli-	bucket, urn, vessel	calpi-
smaragdi-	emerald	smaragdo-
soli-	single	haplo-
soli-	sun	helio-
soli-	alone	oio-
soli-	soil	pedo-
solidi-	firm, solid	stereo-
solidi-	firm, solid	steripho-
solitudini-	isolation	apomono-
solventi-	loosing	lysi-
soni-	sound	phono-, -phone
sordidi-	dirty	pinaro-
sordidi-	dirty, filthy, shabby, soiled	rhyparo-
sorici-	shrew	hyraco-
spadicei-	brown	spadico-
spadoni-	eunuch	thladia-
sparsi-	scattered	scedasto-
sparsi-	scattered	sporadico-
speculi-	mirror	enoptro-
speculi-	mirror	esoptro-
spici-	corn-ear	antherico-
spici-	corn-ear, spike	athero-
spici-, -spica	ear of corn, spike	stachyo-, -stachys
spina-	backbone	rhachi-

spini-	spine	acantho-
sporti-	basket	spyrido-
spuenti-	spitting, a	ptysi-
sputi-	spittle	ptysmato-
squami-	fish-scale	lepido-
stabuli-, tegmini-	stable, stall, shelter	stathmo-
stagnali-	pond	limno-
stagni-	pond, pool	tipho-
stamini-, -stamineus	stamen, thread	stemono-, -stemon
stanni-	tin	cassitero-
stellati-	-starred,starry	astero-
stelli-	star	astro-
stemmati-	pedigree	genea-
stercor-	dung	scybalo-
stercori-	dung	copro-
sterili-	barren	steiro-
sternuti-	sneeze	ptairo-, ptaero-
stillicidii-	drip, drop	stalagmato-
stimulanti-, urenti-	sting	dako-, -dakos
stimuli-	goad	centemato-
stipiti-, stirpi-	stem, stump, block	stypo-
stirpi-	treetrunk	cormo-
storei-, tegeti-	mat	phormo-
straguli-, tapeti-	carpet, rug	tapeto-
straguli-, tapeti-	carpet, rug	tapido-
straguli-, tegmini-, vesti-	covering, quilt	stromne-
straminei-	straw	carpho-
stranguli-	strangle	-anche
strepiti-	noisy	psopho-
strepitu-	noisy	spharago-
strigi-	owl	strigo-
strumosi-	bulky, swollen	onco-, -oncodes
struthioni-	ostrich	strutho-
suavi-	fragrant	thyo-, -thyodes
sub-	up	ana-, ano-
sub-	below, under	hypo-
suberi-	cork	phello-
successori-	follower	amoibo-
succi-	juice, moisture, sap	chylo-, chymo-
succi-	juice, sap	opo-
succidi-, sucidi-	juice, sap	cecido-
sucini-	amber	electro-
-sudes, -vallus	stake	characo-, -charax
suffocati-	choked	pnicto-
suffocati-	choked	strango-
sui-	sow	hyo-, syo-
sulcati-	furrowed	aulaco-
sulci-, -sulcus	furrow	-aulax
summi-	uppermost	hypato-
super-	upon	epi-
super-, supra-	above, over	hyper-
supercili-	eyebrow	ophryo-, -ophrys

supercilii-	eyebrow	episkynio-
supini-	laid back	hyptio-
surculi-	scion, shoot, twig	oscho-, -osche
surculi-	shoot	blasto-, -blastos
surculi-	shoot	clono-
surculi-	shoot	ormeno-
surculi-	shoot, sucker	ptortho-
susurri-	whispering	psithyro-
suti-	stitched	cesto-
suti-	sewn	rhammato-
suti-	sewn, stitched	rhapso-, rhapto-
suturi-	seam	rhaphi-
sylvestri-	wild	agrio-
sylvestri-	forest	hylaeo-, hyleo-, hylo-
tabula-	tablet	schede-
tabuli-	tablet	pinaco-, -pinax
tabuli-	slab, tablet	placo-
tabuli-	board, plank	sanido-
taciti-	quiet	hesychio-
tali-	ankle	sphyro-
talpi-	mole	aspalaco-
talpi-	mole	scalopo-
talpi-	mole	spalaco-
tectati-	covered	scepano-
tecti-	covered	calypso-, calypto-, -calyptus
tecti-	covered	crypto-
tecti-	roof	orophe-
tecti-	shelter	-scepasma
tecti-	shelter	stegano-, -stegia, stego-
-tegens	covering	-bates
teguli-	roofing	erepsi-
telluri-, terra-	earth	chthono-
telluri-, terrestri-	land	geo-
tempori-	time	chrono-
tendicula-	trap, part of a	schastero-
tenebri-	darkness	scoto-
tenebri-	darkness	zopho-
tenenti-	holding	hapto-
teneri-	delicate, tender	atalo-
teneri-	delicate	habro-
teneri-	delicate	hapalo-
tensi-, tenti-	stretched	teino-, tino-
tenui-	thin	lagaro-
tenui-	thin	lepto-
terebra-	auger, gimlet	trypano-
terebrati-	bored	trypeto-
terra-	dry land	cherso-
terribili-	frightful	gorgo-
testi-	testicle	orchi-
testudini-	tortoise	chelono-
testudini-, tholi-	dome	tholo-

textili-	web, woven	plocio-, ploco-, -ploca, -ploce
tibii-	flute	aulo-
tibii-	shin	cneme-
tigri-	tiger	tigro-
tincti-	dyed	bapho-, bapsi-
tincti-	dyed, stained, tinged	chroso-, chrozo-
tinei-	moth	phalaino-
tonitru-	thunder	bronto-
tori-	muscle	ino-
tori-	quoit	disco-
torrenti-	mountain-stream	charadro-
torrenti-	stream	rhyaco-
torti-	plaiting, twisting, weaving	plegmato-, -plegma, plexi-
torti-	twisted	streblo-, stremmato-
torti-	twisted	strepho-, strepsi-, strepto-
torti-	twisted	stropho-
toxico-	arrow-poison	toxico-
trabi-	beam	trapheco-
trans-	across	peran-
tremuli-	quiver	einosi-
triangulari-	triangular	delto-
tribu-	tribe	phylo-
tribuli-	caltrops	tribolo-
tridenti-	trident	thrinaco-, -thrinax
triquetri-	3-cornered	trigono-
tristitii-	gloom	stygno-
triti-	rubbed, worn	tribaco-, tribo-
tritici-	wheat	pyro-
trulli-	ladle	arytero-, arytaino-
truncati-	cut short, docked	colousto-
tuberi-	truffle	hydno-
tubi-	pipe	ocheto-
tubi-	pipe	siphono-, -siphon
tubi-	pipe	soleno-, -solen
tubi-	pipe	syringo-, -syrinx
tumidi-	swollen	cymato-
tumidi-	swollen	oedo-
tumuli-	mound	tymbo-
tunica-	tunic	-chiton
turba-	crowd	ochlo-
turbati-	disturbed, troubled	tarakto-
turbini-	top	rhombo-
turbini-	top	strombo-
turri-	tower	pyrgo-
turri-	tower	tyrsio-
tympani-	drum	tympano-
typi-	blow, copy, figure, image	typo-
typi-	impression, mode	typo-
typi-	print, replica	typo-

-ugo	(substantival suffix attached to adjectives)	-osyne, -otes
ultimi-	last	eschato-
umbelli-	parasol	sciado-
umbilici-	navel	omphalo-
umboni-	shield-boss, protuberance	exoncomato-, exoncosi-
umbra-	shade, shadow	scia-, scio-
umbrati-	shaded	scioto-, -sciodes
uncini-	claw	-chele
ungui-	claw	onycho-
unguli-	hoof	hople-
uni-	single	mono-
urbi-	city	asti-, asty-
urcei-, urceoli-	jug, pitcher	cyrillio-
urenti-	burning, caustic, corrosive	caustico-
ursi-	bear	arcto-
ustulati-	burnt	aetho-
uteri-	womb	coeleo-, coelia-, coelio-
uteri-	womb	delphy-, delphyo-
uteri-	womb	hystera-
uteri-	womb	koilia-, koilio-
uva-	bunch of grapes	staphylo-
vacui-	empty	ceno-
vadi-	shallows, shoal	tenago-
vagini-	sheath	coleo-
validi-, robusti-	strong, stout	alkimo-
validi-	strong	iphi-
validi-	strong	ischyro-
validi-	strong	sterrho-
vanni-	winnowing-fan	likno-
vanni-	winnowing-fan, -shovel	ptyo-
variegati-	many-coloured	poecilo-, poikilo-
vasi-	vessel	angio-, ango-
veli-	veil	calyptro-
velleri-	fleece	poco-
velli-	fleece	naco-
vena-	vein	phlebo-, -phleps
veneni-	poison	ios-
veneni-	poison	pharmaco-, pharmako-
venti-	wind	anemo-
venti-	wind	pneumato-
ventri-	belly	gastro-
venusti-	beauty, grace	-charis
veri-	truth	aletho-
veri-, verni-, vernali-	Spring	earo-
vermi-	worm	helmintho-
verruci-	wart	phymato-
versi-	sweep	saro-
versi-	change, turn	tropo-, -trope
vertebra-	backbone	sphondylo-
veru-	spit	obelo-

vesici-, vesiculi-	bladder	asco-, cysti-
vespa-	wasp	spheco-
vesperi-	evening	hespero-
vespertili-, vesperugini-	bat	nycteri-
vesti-	clothing	esthe-
veteri-	old, ancient	archaeo-, archaio-
veteri-, vetus-	old	palaeo-
vexilli-	standard	semeio-, -semeia
via-	way	hodo-
vicini-	neighbour	-geiton, -geton, -giton
villosi-	shaggy	lasio-
vimini-	willow-twig	lygo-
vimini-, virga-	wand	thyrso-
vini-	wine	oeno-
viola-	violet	io-
violacei-	violet-coloured	ianthino-
viperi-	adder	echidno-, echi-
virelli-, viriduli-	greenish	chloano-
virga-	rod	canono-
virga-	shoot, twig	rhadamno-
virgini-	maiden	partheno-
viridi-	green	chloro-
viscera-	bowels, entrails, gut	splanchno-
visci-	sticky	ixod-
vita-	life	zoe-
vitri-	glass	hyalo-
vittati-	striped	rhabdoto-
vituli-	calf	moscho-
-vivus	living	-bius
voluti-	coiled	rhymbo-
voluti-	coil	speiro-
vomeri-	ploughshare	hyno-
vortici-	eddy, whirl	strobo-
vulpi-	fox	alopeco-

ENGLISH	LATIN	GREEK
(diminutives)	-ellus, -illus, -ulus	-idion
(substantival suffixes attached to adjectives)	-ugo	-osyne, -otes
abnormal, irregular	enormi-	anomalo-
above, over	super-, supra-	hyper-
abundant, copious	copiosi-	perisso-
abyss, gulf	gurgiti-, profundi-, voragini-	barathro-
acorn	glandi-	balano-
across	trans-	peran-
active, vigorous	efficaci-	drastico-
acute	acuti-	oxy-
adder	viperi-	echidno-, echi-
after	post-	meta-
again	re-	ana-
again, back	re-	palin-
agreeable, pleasant	amoeni-	terpno-, terpsi-
agreeable, pleasant	amoeni-, grati-, iucundi-	hedy-
air	aeri-	meteoro-
all	omni-	pan-, panto-
alone	soli-	oio-
altar	ara-	thymeli-
always	semper-	aei-, ai-
amber	sucini-	electro-
among	inter-	meta-
ample	ampli-	perisso-
anchor	ancora-	ankyro-
angle	anguli-	-gonia
angled	angulati-	gonio-
animal	animali-	zoo-
ankle	tali-	sphyro-
ant	formica-	myrmeco-
ant's-nest	grumuli-	myrmedono-
anvil	incudi-	acmo-
aperture	foramini-	-trema
appearance	faciei-	-opsis
apple	mali-	melo-
aquatic	aquatici-, aquatili-	enydro-, enygro-
arm	brachii-	brachio-
armed	armati-	enoplo-
armful	manipuli-	angkalido-
armpit	axilli-	maschalo-
around	circum-	peri-
arrangement	collocati-, dispositi-	taxi-
arrow	sagitti-	belo-
arrow	sagitti-	toxeumato-
arrow-poison	toxico-	toxico-
ascending	ascendenti-, acclivi-	anophero-
ash-grey	cinerei-	spodio-
ash-grey	cinerei-	tephro-
ashes	cineri-	spodo-

ass	asini-	ono-
auger, gimlet	terebra-	trypano-
autumn	autumnali-	phthinoporo-
autumnal	autumnali-	metoporino-
away from	ab-	apo-
away from, beyond, out of, without	extra-, ultra-	exo-
axe	securi-	axino-
axe	securi-	peleceo-, pelecy-, pelyco-
axle	axi-	axono-
back	dorsi-, tergi-	noto-
backbone	spina-	rhachi-
backbone	vertebra-	sphondylo-
bad	male-	caco-
bad, evil	male-, mali-	phaulo-
bad, ill	male-	dys-
bag	sacci-	phascolo-
bag, pouch, sack	culei-, folli-, folliculi-	thylaco-, -thylax
bag, pouch, sack	marsupi-, sacci-	thylaco-, -thylax
bag, pouch, sack	sacculi-, uteri-	thylaco-, -thylax
bag, sack	sacci-	sacco-
bald	calvi-, glabri-	phalacro-
bald	calvi-, glabri-	psedno-
baldric, belt	baltei-	zostero-
ball	globi-	sphaero-
ball of wool	glomeri-	tolypo-
band	ligamenti-	sphinctero-
band	ligamenti-	zeugmato-
band	liguli-, lori-	desmo-, desmos
band, belt, strap	baltei-, cinguli-, lori-, zoni-	telamono-
band, strap	fasciari-, liguli-, lori-	taenio-
baneful, damaging, harmful, hurtful, injurious	nocenti-, nocivi-, noxii-	blabero-, blabo-
baneful, damaging, harmful, hurtful, injurious	nocenti-, nocivi-, noxii-	blapto-, blaptiko-, -blapton
bank, mound, hillock	aggeri-, tumuli-	ochtho-
bar	sera-	rhabdo-
bare	calvi-, glabri-	psilo-
bark	cortici-	phloeo-, phloio-
barren	sterili-	steiro-
base, bottom	fundi-	pyndaco-
base, bottom	fundi-	pythmeno-
base, pedestal	basi-	bathro-
basket	corbi-	calatho-, -calathus
basket	corbi-	larco-
basket	sporti-	spyrido-
bastard	nothi-	notho-
bat	vespertili-, vesperugini-	nycteri-
beak	rostri-	rhamphi-, rhampho-
beak, snout	rostri-	rhyncho-

beam	trabi-	trapheco-
bear	ursi-	arcto-
beard	barba-	geneiado-
beard	barba-	pogon(o)-, -pogon
bearing, carrying	-fer, -ger	-phoros
beaten, bruised, pounded, struck	contusi-	copto-
beautiful, charming, lovely	pulchri-, venusti-	calli-, calo-
beauty, grace	venusti-	-charis
beaver	fibri-	castoro-
bed, bedding, layer	lamelli-, strati-	stromato-
bed, couch, mattress	cubili-, lecti-, lectuli-	stromne-
bedbug	cimici-	corio-
bedchamber	cubiculi-	domatio-
bedchamber	cubiculi-	thalamo-
bee	api-	melisso-, melitto-
beetle	scarabaei-	cantharo-
before, in front of	prae-	pro-, proso-
behind, back	posteri-	opistho-
being, essence, existence	essenti-, natura-, vita-	hyparxeo-
being, essence, existence	essenti-, natura-, vita-	onto-
bell	campani-	codono-, -codon
bellowing, roaring	fremitu-, mugitu-, strepitu-	bryche-, brycheto-
bellows	folli-	physo-
belly	ventri-	gastro-
below, down	infra-, sub-	cata-
below, under	sub-	hypo-
belt, girdle	cinguli-	zono-
belt, girdle	cinguli-, zoni-	stelmono-
bend	flexu-	ankono-
bending, a	flexi-	campe-
bent	flecti-	campso-, campto-, campylo-
bent	flexi-	scolio-
bent	flexi-, flecti-	rhoiko-
bent arm	flectibrachii-	angkale-
bent, crooked	flexi-	scambo-
bent, crooked	flexi-, obstipi-	rhaibo-
bent, crooked	pravi-	ankylo-
bent, crooked, curved	curv(at)i-, pandi-, pravi-	coronido-
berry	bacca-, -coccus	coccy-, -coccos
berry	bacci-	rhago-
beside	iuxta-	para-
besom, broom	scopae-	corema-
besom, broom	scopae-	corethro-
best	optimi-	aristo-
bile	bili-	chole-, cholo-
billhook	falculi-	copido-
billow, swell, wave	fluctu-	cymo-
bird	avi-	orneo-, ornitho-
bit, piece	frusti-	thrymmato-
bite	morsu-	dako-, -dakos

bitter	amari-	ateramno-
bitter	amari-	picro-
black	nigri-	celaeno-, celaino-
black, dark	atro-	pelio-, pelidno-
black, shiny	nigri-	melancho-, melano-
bladder	vesici-, vesiculi-	asco-, cysti-
blade	lamina-	spatho-
blind	caeci-	typhlo-
blister, pustule, vesicle	pustuli-, vesici-	phlyctaino-
blister, pustule, vesicle	pustuli-, vesici-, vesiculi-	psydraco-
blistered, blown out, bubbled	bullati-	pemphigo-
bloody	cruenti-, sanguinei-	haemato-
blow, copy, figure, image, impression, mode, print, replica	typi-	typo-
blown out	inflati-	physeto-
blown out	inflati-	pneumatico-
blue-grey	glauci-	glauco-
blunt	obtusi-	ambly-
blunt	obtusi-	copho-
board, plank	tabuli-	sanido-
boat	cymba-	cymbi-
boat	lintri-, naviculi-	scapho-
body	corpori-	somato-
bold	audaci-	thrasy-
bolster, cushion, mattress, pillow, quilt	culciti-	tyle-, tyleio-
bolt	obici-, pessuli-	cleithro-
bolt	repaguli-, sera-	cleithro-
bolt	obici-, pessuli-	gompho-
bolt	repaguli-, sera-	gompho-
bone	ossi-	osteo-, osto-
boot, shoe	calcei-, caliga-, cothurni-	crepido-
border	margini-	craspedo-
bored	excavati-, perforati-, terebrati-	trypeto-
bosom	sinu-	colpo-
bound	ligati-, stricti-, vincti-	sphigmato-
bound	ligati-, stricti-, vincti-	sphincto-
boundary	fini-	horio-, horismo-
bow	arci-	toxo-
bowels, entrails, gut	viscera-	splanchno-
bowl	crateri-	cratero-
bowl	crateri-	lecano-
bowl	crateri-	scaphido-
bowl	crateri-	tryblio-
box	arca-	larnaco-
box	arca-, capsa-, cista-, pyxi-	ciboto-
box	capsi-	pyxido-, -pyxis
box, chest	cista-	cisto-
box, container	alveoli-, capsula-	cypseli-
brain	cerebri-	encephalo-

bran	furfuri-	pityro-
branch	rami-	clado-
branch	rami-	rhadico-
branch	rami-	spadico-
branch	rami-	thallo-
branchlets	ramuli-	cladio-, cladisco-
bread	pani-	arto-
bread	pani-	sitio-
break, fracture, laceration, rent, rupture, tear	rima-	regmato-
breakable, brittle	fragili-	thraulo-, thrausto-
breast	mammi-	masto-
breast	pectori-	sterno-
breastplate	lorica-	thoraco-
breath, life, soul	animi-	psycho-
breeze	aura-	aura-
brick	lateri-	plintho-
bridge	ponti-	gephyro-
bright	clari-	liparo-
bright	nitidi-	gano-
bright	nitidi-	lampro-
bringing	ferenti-	phere-
bristle	seti-	chaeto-, -chaete
bristling	horridi-	phrisso-, phrixo-
brittle	fragili-	acampo-
brittle	fragili-	clao-
brittle, crumbly, loose	fragili-, friabili-	psathyro-
broad, wide	lati-	eury-
broad, wide	lati-	platy-
broadsword	gladii-	rhomphaio-
broken	fracti-	clasto-
broken	fracti-	klasto-
broken	fracti-	thrypsi-, thrypto-
broken, cut off	fracti-, succisi-	trypho-
bronze, copper	aeri-	chalco-
broom	scopi-	saro-
brother	fratri-	adelpho-
brown	spadicei-	spadico-
brush	penicilli-	callyntro-
bucket, urn, vessel	situli-	calpi-
bugle, trumpet	buccina-	salpingo-, -salpinx
bulb	bulbi-	bolbo-
bulky, stout, thick	corpulenti-, crassi-, obesi-	hadro-
bulky, swollen	strumosi-	onco-, -oncodes
bull	bovi-	tauro-
bunch of grapes	uva-	staphylo-
bundle	fasci-	angkalo-
bundle	fasci-	-desme
bundle	fasci-	tropalo-
bundle, fagot	fasci-, fasciculi-	phacelo-
bung, plug, stopper	obturaculi-, obturamenti-	bysmo-, bystro-
bung, plug, stopper	obturatori-	bysmo-, bystro-

burial	funeri-, sepultura-	tapho-
burning, caustic, corrosive	urenti-	caustico-
burnt	cremati-, incensi-, tosti-, usti-	causto-
burnt	ustulati-	aetho-
burrow, dig	fodi-	orycho-
burrowed, dug	fossi-	orycto-
butter	butyri-	butyro-
cable	funi-	sparto-
calf	vituli-	moscho-
callus, knob, knot, lump	bulla-, glaeba-, nodi-	tylo-
caltrops	tribuli-	tribolo-
camel	cameli-	camelo-
cane, club, cudgel, staff	baculi-, bacilli-	bactero-, bactro-
carmine, crimson	punicei-	phoeniceo-, phoenico-
carpet, rug	straguli-, tapeti-	tapeto-
carpet, rug	straguli-, tapeti-	tapido-
carved	sculpti-	glypho-, glypto-
case	-capsa	-theca
cat	feli-	ailuro-
catgut	chorda-	chorde-, chordo-
cause	causa-	aetio-
cave	antri-, cavi-, caverni-	antro-
cave	specu-, spelunci-	antro-
cave	caverni-	spelaio-
celebrated	celebri-	-cles
cell	celluli-	cyto-
chaff	palea-	achyro-
chain	cateni-	halysi-
chalice	calici-	depao-, -depas
chalk	creta-	titano-
change	mutationi-	allago-
change, turn	versi-	tropo-, -trope
changed	mutati-	meta-
chasm	hiatu-	chasmato-
chasm	hiatu-	sarmato-
chasm, chink, cleft	rima-	regmato-
chasm, cleft	chasmati-, fissuri-, hiati-, rimi-	barathro-
chasm, cleft, gully, ravine	fissura-, hiatu-, rima-	pharango-
cheat, impostor	fraudi-	phenaco-, -phenax
cheese	casei-	tyro-
chest	pectori-	stetho-
chief	principi-	arche-, archi-
chin	menti-	geneio-
chink, crack, rent	rima-	rhagado-
choked	suffocati-	pnicto-
choked	suffocati-	strango-
chopper, cleaver	dolabri-	copido-
cicada	cicada-	tettigo-
city	urbi-	asti-, asty-
clamour, din, noise	crepitu-, fremitu-, strepitu-	celado-
class	classi-	phylo-

claw	uncini-	-chele
claw	ungui-	onycho-
clay	argilla-	argillo-
clay	argilla-, creta-	ceramo-
clay	luti-	pelo-
clean, pure	casti-, puri-	catharo-
cleansing, purgative	purgativi-	cathartico-
cliff	scopuli-	cremno-
climbing	scandenti-	periallocaulo-
clinging	amplecti-	aspazo-
cloak	(no Latin equivalent)	chlamydo-
cloak	palliati-	-chlaena
close, crowded	aggregati-, coacervati-	adino-
close, crowded	coarctati-, conferti-, congesti-	adino-
close, crowded	conglomerati-, crebri-	adino-
close-packed	conferti-	araro-
close-pressed	compacti-	stipho-, stiphro-, stipto-
closed, shut	clausi-	cleisto-
clot, lump	grumi-	thrombo-
clothing	vesti-	esthe-
cloud	nubi-	nepheo-, nepho-, nephelio-
club	clavi-	cordyle-
club	clavi-	coryne-, coryno-, -coryne
club	clavi-	rhopalo-
cluster	racemi-, uvi-	botryo-, -botrys
coarse	rudi-	trachy-
cock	galli-	alectoro-, alectryo-
coil	voluti-	speiro-
coiled	voluti-	rhymbo-
colander, sieve, strainer	cribri-	ethmo-
cold	frigidi-	psychro-
colour	-color	-chroma, -chroos
coloured	colorati-	chromato-
colt, foal, pony	equulei-, equuli-	hippario-
comb	pectini-	cteno-
common	communi-, vulgari-	coeno-
cone	coni-	cono-
confused, disordered	confusi-	asystasio-, asystato-
constricting	astringenti-	stryphno-
constricting	astringenti-	styphelo-, stypho-
constricting	astringenti-	stypsi-, styptico-
constriction	constrictioni-	sphinxi-
coot	fulici-	phalaro-
coppice, copse, thicket	dumeti-, fruticeti-, silvuli-	drymo-
cord, string	funiculi-, resticuli-	thomingo-, -thominx
cork	suberi-	phello-
corn	frumenti-	sito-
corn-ear	spici-	antherico-
corn-ear, spike	spici-	athero-
cornice, eaves	protecti-, sima-	geiso-, geisso-
couch	cubili-	clino-, -cline
countless	innumeri-	myrio-

cover	operculi-	elytro-
cover	operculi-	pomato-, pomatio-
covered	tectati-	scepano-
covered	tecti-	calypso-, calypto-, -calyptus
covered	tecti-	crypto-
covering	involucri-	eilemato-
covering	-tegens	-bates
covering, a	integumenti-, involucri-	-calymma
covering, a	operculi-	-calymma
covering, quilt	straguli-, tegmini-, vesti-	stromne-
cowl	cuculli-	(no Gr. equiv.)
crab	cancri-	carcino-
crab	cancri-	grapsaio-
cracked, split	fissi-	rhexi-
crackling, noisy, roaring	fremitu-	bromeso-
crane	grui-	gerano-
creeping	repenti-	herpesti-
crescentic	lunati-	menoidi-
crest	crist(at)i-	corytho-
crest	crista-	lopho-
crest, tuft	crista-	crobylo-
crooked	anfractuosi-	ancyclo-
crooked	pravi-	cyllo-
cross	cruci-	stauro-
crosswise	decussati-	chiasmato-
crow	cornici-	corone-
crowd	turba-	ochlo-
crowded	conferti-	athro-
crowded	frequenti-	thamino-
crown	corona-	-stemma
crown	corona-	stephano-, -stephus
crown, garland, wreath	serti-	-stelma
crumb, morsel	frustuli-, mica-	psacado-
crumb, morsel	frustuli-, mica-	psicho-
crumb, morsel	frustuli-, mica-	psom(i)o-
cube	cubi-	cubo-
cuckoo	cuculi-	coccygo-
cudgel	baculi-, fusti-	cordyle-
cudgel	fusti-	scytalo-
cup	acetabuli-	cyatho-
cup	calici-, poculi-, scyphi-	cylico-
cup	cupuli-	-calathus
cup	cupuli-	cotylo-, -cotyle
cup	cupuli-	scypho-
curl	cincinni-	bostrycho-
curl	cincinni-, cirri-	eligmato-
curl	cincinni-, cirri-	heligma-
curl	cincinni-, cirri-	plocio-, ploco-, -ploca, -ploce
curtailed, shortened, stunted	curti-	colobo-

curved	curvati-	cyrto-
cut	secti-	temno-
cut	secti-	tmesi-, tmeto-
cut	serrati-	prismato-, pristo-
cut short, docked	truncati-	colousto-
cut, piece, slice	secti-	tomo-
cuttlefish	sepia-	teutho-
damp, moist, wet	humidi-, made-	notero-
damson, plum	pruni-	coccymelo-
dappled	guttati-, maculosi-	psaro-
dark blue	caerulei-	cyano-
dark, dusky	fusci-	phaeo-
darkness	tenebri-	scoto-
darkness	tenebri-	zopho-
dart	iaculi-, teli-	belo-
dart	jaculi-, teli-	akonto-
dawn	aurora-, diluculi-	orthrio-, orthro-
dawn	aurori-	eo-
day	diurni-	hemero-
dead	morti-	necro-
death	morti-	thanato-, -thanasia
decay	putridi-	-phthora
decaying	putrescenti-	maranto-
deep	alti-, profundi-	bathy-
deep (the), depth	alti-, altitudini-	bysso-, bytho-
deep (the), depth	ponti-, profundi-	bysso-, bytho-
deep red	rubidi-	pyrrho-
deep yellow	lutei-	xantho-
delicate	teneri-	habro-
delicate	teneri-	hapalo-
delicate, tender	teneri-	atalo-
delightful, sweet	dulci-, suavi-	hedy-
delighting in, rejoicing in	laeti-	chaero-
dense	densi-	pycno-
desert	deserti-	eremo-
despised	contempti-, despicati-	atimeto-
despised	humili-, spreti-	atimeto-
dew	rori-	droso-
different	differenti-	diaphoro-
different	dissimili-	hetero-
dipped	mersi-	bapti-
dirt	coeni-, caeni-	borboro-
dirty	sordidi-	pinaro-
dirty, filthy, shabby, soiled	sordidi-	rhyparo-
diseased	aegri-, morbidi-	astheno-
dish	catilli-, catini-	lopado-
distant	remoti-	macryno-
disturbed, troubled	turbati-	tarakto-
ditch	fossi-	taphro-
diving	mersi-	colymbi-
doctor	medic-	iatro-

dog	cani-	cyno-
dolphin	delphini-	delphino-
dome	testudini-, tholi-	tholo-
door	ianua-, -ianua	thyra-, -thyra
double	ancipiti-, duplici-	diplo-
double	duplici-	disso-, ditto-
doubled, folded	duplicati-	diptycho-
down	lanugini-	chnoo-
down	lanugini-	lachno-
dragon	draconi-	draconto-
drawn	delineati-	grapho-, grapto-
dreadful	diri-, terribili-	deino-
dregs, lees, sediment	amurca-, faeci-	trygo-
drip, drop	stillicidii-	stalagmato-
driven away	expulsi-	elaterio-
drone	fuci-	cepheno-
drop	gutta-	stagono-
drum	tympani-	tympano-
dry	aridi-, sicci-	azaleo-
dry	aridi-, sicci-	xero-
dry land	terra-	cherso-
duck	anati-	nesso-, netto-
dug	fossili-	scapto-
dull black	atri-	melancho-, melano-
dung	fimi-	ontho-
dung	fimi-	scato-
dung	stercor-	scybalo-
dung	stercori-	copro-
dusky	fusci-	orphno-
dusky	fusci-	percno-
dusky	fusci-	zophero-
dust	pulveri-	coni-, conio-, conis-
dust	pulveri-	psocho-
dwarf	pumili-	nano-
dwelling in	incolenti-	endemo-
dyed	tincti-	bapho-, bapsi-
dyed, stained, tinged	tincti-	chroso-, chrozo-
each, every	(quisque)	ecasto-
eagle	aquili-	aeto-
ear	auri-	oto-
ear of corn, spike	spici-, -spica	stachyo-, -stachys
earlier	praecociori-	protero-
early	praecoci-	proio-
earth	telluri-, terra-	chthono-
earthworm	lumbrici-	scoleco-
east	orienti-	anatoli-
eatable	eduli-, esculenti-	brosimo-
eating	esu-	phago-
eddy, whirl	vortici-	strobo-
edge	margini-	ityo-
eel	anguilla-	enchelyo-

egg	ovo-	oo-
elbow	cubiti-	olene-, oleno-, olecrano-
embrace	amplexi-	aspasio-
emerald	smaragdi-	smaragdo-
empty	vacui-	ceno-
end	fini-, ultimi-	teleio-, teleo-, telo-
end	fini-, ultimi-	teleuto-
endowed	praediti-	proikto-
entire, whole	integri-	holo-
equal	aequi-	iso-
eunuch	spadoni-	thladia-
even	aequi-, pari-	artio-
even, level	plani-	homalo-
evening	vesperi-	hespero-
ever-	semper-	empedo-
evergreen	sempervirenti-	empedophyllo-
everlasting, undying	aeterni-, immortali	athanasio-, athanato-
everlasting, undying	sempiterni-	athanasio-, athanato-
evident	manifesti-	delo-
evident	manifesti-	phanero-
excessive	nimii-	perisso-
exchange	alterni-	enallago-
exchanging	permutanti-	amoibo-
exudate, ooze	gutta-	stacto-
eye	oculi-	blemmato-, -blemma
eye	oculi-	ophthalmo-
eyebrow	supercili-	ophryo-, -ophrys
eyebrow	supercilii-	episkynio-
eyelash	cilii-	blephari-, blepharido-
eyelet	ocelli-	ommato-, -omma
eyelid	palpebri-	blepharo-, -blepharon
face	personati-	-ops
fading, withering	deflorescenti-, marcescenti-	maranto-
falcon, harrier, hawk	falconi-	circo-
fall	caduci-	pipto-
fall	caduci-	ptoseo-, -ptosis
fall, slip	lapsu-	sphalmato-
fallen	casi-	ptoto-
false, spurious	falsi-, spurii-	pseudo-
fan	flabelli-	rhipi-, rhipidi-
far	longe-	tele-
fat, grease	adipi-, pingui-	piaro-
fat, grease	adipi-, pingui-	pimeleo-, pimelo-
fawn	hinnulei-	elapho-
fawn	hinnulei-	kemado-
fawn	hinnulei-	nebro-
fear-generating	formidabili-	phobero-
feather	penni-, plumi-	ptilo-
feathered	pennati-, plumati-, plumosi-	pteno-
feathered	pennati-, plumati-, plumosi-	pteridio-, pterino-
felloe	curvatura-	sotro-

felt	coacti-	pileo-, pilo-
female	foeminei-	gyno-
female	foeminei-	thely-
fence	septi-	erymno-
fence, hedge, partition	septi-	phragmo-, phragmato-, -phragma
fern	filici-	pteri-, pterido-
few	pauci-	oligo-
few	pauci-	pauro-
few	pauci-	spano-
fibre, ligament, nerve, sinew, tendon	nervi-	neuro-
fibre, sinew	fibra-, nervi-	inio-
fibrous, sinewy	fibrati-, nervosi-	-inodes
field	arvi-	arouro-
field, country	agresti-	agro-
fiery red	flammei-, ignei-	pyrso-
fig	fici-	ischado-
fig	fici-	syco-
figure, form, shape	forma-, specie-	eido-
figure, form, shape	habitu-	schemato-
file	limi-, scalpri-	xystero-
filings, shavings	ramenti-, scobi-, stringenti-	xysmato-
filtered, strained	percolati-	hylisto-
-finch	-fringilla	-spiza
fine linen, muslin	byssi-	sindono-
finger	digiti-	dactylo-
finished, polished, smooth	politi-	glaphyro-
fire	flagr-, flammi-, igni-, incensi-	pyro-
firm, lasting, stout	firmi-, solidi-, stabili-	ochyro-
firm, solid	solidi-	stereo-
firm, solid	solidi-	steripho-
first	primi-	proto-
fish	pisci-	ichthyo-
fish-hook	hamuli-	ancistro-
fish-scale	squami-	lepido-
fishing-net, -basket	reti-	gripho-
fixed, made fast, secured	-fixus, -fixa	pegmato-, -pegia
flabby	flaccidi-	chauno-
flagon, flask	ampulla-, lagena-, laguncula-	lageno-, lagyno-
flame	igni-, flamma-	phlogi-, -phlox
flask	ampulli-	lecytho-
flat	plani-	petalo-
flat	plani-	placo-
flat dish, patella	-patina, -patella	-patane
flax, linen	lini-, lintei-	othono-, othonio-
flea	pulici-	psyllo-
fleece	velleri-	poco-
fleece	velli-	naco-
flesh	carni-	crea-, creio-, creo-
flesh	carni-	sarco-
flesh, meat	carni-	broto-

80

floury, mealy	farini-	aleuro-
flow	fluxi-, flucti-	rheo-
flow	fluxi-, flucti-	rheumato-
flow	fluxi-, flucti-	-rrhoe, -rrhoea, -rrhoa
flower	flori-	antho-, -anthemon
flower-cup	(no Latin equivalent)	calyco-
-flowered	-flora	-antha, -e
flowery	florei-, floridi-	antheio-
flowing together	confluxi-	syrrheo-
fluid, juice	seri-, sero-	ichoro-
flute	tibii-	aulo-
fly	musci-	myio-, -myia
fold	plica-	ptysso-
folded	plicati-	ptycho-
folded	plicati-	ptygmato-
folded	plicati-	ptykto-
folding, a	plicanti-	ptyxo-
follower	successori-	amoibo-
food	cibi-, pabuli-	bromato-, -broma
food	cibi-, pabuli-	sitio-
food	nutrimenti-	tropho-, -trophe
food, feed	cibi-, pabulo-, nutrimenti-	trepho-
foot	-pes	-pus, -podos
foot, -footed	pedi-, -pes	podo-, -poda
forearm	cubiti-, ulni-	pecheo-, pechy-
forehead	fronti-	metopo-
foreign, stranger	alieni-, peregrini-	xeno-
forest	sylvestri-	hylaeo-, hyleo-, hylo-
forked	furcati-	dicraeo-
fox	vulpi-	alopeco-
fragment, piece	frusti-	clasmato-
fragrant	fragranti-	euod-, euosm-
fragrant	fragranti-	myristico-
fragrant	suavi-	thyo-, -thyodes
free	liberi-	eleuthero-
frightful	terribili-	gorgo-
fringe	fimbria-	cosymbo-
fringe	fimbri-	lomato-, -loma
fringe	fimbri-	thysano-
fringed	fimbriati-	crossoto-
frog	rani-	batracho-
frost	gelu-	cryo-, crymo-
frost	gelu-	rhigo-, rhigio-
frost	gelu-, geli-	pageto-
fruit	frugi-	carpo-
-fruited	-fructa	-carpa
fruitstone	putamini-	pyreno-
full	pleni-	pleo-, plero-
fungus	fungi-	myceto-, myco-, -myces
funnel	infundibuli-	choano-, chono-
furrow	sulci-, -sulcus	-aulax
furrowed	sulcati-	aulaco-

gable, pinnacle, roof	fastigi-	pterygio-
gadfly	asili-, tabani-	oistro-
gall	galli-	cecidio-
gaping	hianti-	chaeno-
gaping	hianti-	chasco-
garden	horti-	cepo-
garland, wreath	serti-	-stemma
garland, wreath	serti-	stephano-, -stephus
garlic	allii-	scorodo-
gate	-foris, -porta	pylo-, -pyle
gift	donati-	doro-
girdled	cincti-	zomato-
girdled	cincti-	zosto-
glade	nemori-	nemeo-
glance, look	aspectu-	blemmato-, -blemma
gland	glanduli-	adeno-
glass	vitri-	hyalo-
gloaming	diluculi-	lycophoto-
gloom	tristitii-	stygno-
glory	gloriosi-	-doxa
glossy, polished, shining	nitenti-, splendenti-	stilbo-, stilpno-, stilpsi-
glue	glutini-	colla-, collo-
glue	glutinosi-	gloeo-
gnat	culici-	conopso-
gnawed	rosi-	trocto-
gnawing	morsu-, rodenti-	troxi-
goad	stimuli-	centemato-
goad, quill, spur, sting	calcari-	plectro-
goad, spur	calcari-	kentro-, -kentron
goat	capra-, capri-	aego-, aegi-
goatskin	hircipelli-	sisyro-
goblet	poculi-, scyphi-	cypello-
God	Dei-	Theo-
gold	aurei-, aureo-	chryso-
good	boni-	agatho-
goose	anseri-	cheno-
graceful	gracili-	charito-
graft, scion, slip	insiti-, surculi-	clemato-, -clema
grape	acini-	rhago-
grape	acini-	staphylo-
grass	gramini-	-chloe
grasshopper	locusta-	acrido-
gravel, shingle	glarea-	cherado-
great	magni-	mega-
greater	maiori-	meizo-
green	viridi-	chloro-
greenish	virelli-, viriduli-	chloano-
grey, grizzled	cani-, incani-	polio-
griffin	gryphi-	grypo-
ground	humi-, soli-	edapho-
grown together	coaliti-, connati-	symphyo-
growth	auctu-, incrementi-	-phye(s)

growth	creti-	auxi-
guard	custodi-	-teres, -teretes
guard	custodii-	phylaco-, -phylax
guide	duci-	hegemono-
guide	duci-	hodego-
guineafowl	meleagridi-	meleagrido-
gut	intestini-	entero-
hail	grandini-	chalazo-
hair	capilli-, crini-	etheiro-
hair	capilli-, crini-, pili-	come-, como-, -come
hair	coma-, crini-, pili-, villi-	-thrix
hair	coma-, crini-, pili-, villi-	tricho-
hairnet	reticuli-	cecryphalo-
half	semi-	hemi-, hemisy-
hammer	mallei-	sphyra-
hand	manu-	cheiro-, chero-, chiro-
handle	ansi-, capuli-, manubri-	steleo-
hanging	pendenti-, penduli-	cremasto-
hard	duri-	aageo-
hard	duri-	ateramno-
hard	duri-	sclero-
hardened	indurati-	atyloto-
hare	lepori-	lago-
hare	lepori-	ptaco-, ptoco-
harp, lute, lyre	lyra-	cithara-
harrow	irpici-	bolokopo-
harsh	asperi-	ateramno-
hat	petasi-	petaso-
hawk	accipitri-	hieraco-
hay	foeni-	chorto-, -chortos
he-goat	hirci-	trago-
head	capiti-	cephalo-
headdress	mitri-	mitro-
-headed	-capitata, -ceps	-cephala, -e
heap	cumuli-	themono-
heap, mound	cumuli-, grumuli-, tumuli-	soro-
heaped up	coacervati-, congesti-	corysto-
heaped up	cumulati-	corysto-
heaped up, stored up	coacervati-, collecti-	keimelio-
heart	cordi-	cardio-
heaven	caeli-, coeli-	urano-
heavy	gravi-	bary-
heavy	gravi-, ponderosi-	embritho-
hedgehog	erinacei-	chero-
hedgehog	erinacei-	echino-
hedgehog	erinacei-	schyro-
heel	calci-	pterno-, -pterna
helmet	cassidi-, galea-	corytho-
herald	nuntii-	keryko-
heron	ardei-	erodio-
hidden	abditi-, celati-, occulti-	ceutho-

hidden	abditi-, celati-, occulti-	crypto-
hide	corii-, pelli-	phorino-
hide, leather, skin	aluta-, cori-, pelli-	spato-
high	alti-, elati-, proceri-	hypselo-, hypsi-
higher	altiori-	hypsitero-
highest	altissimi-	hypsisto-
hill	clivi-, colli-	colono-
hill	colli-	bouno-
hinge	cardini-	ginglymo-
hinged	cardinali-	ginglymato-
hip	coxi-	ischio-
hoar-frost, rime	pruina-	pachne-
hoe	sarculi-	scalido-
holding	tenenti-	hapto-
holding fast to	firmi-, stabili-	aptosi-, aptoto-
hole	foramini-	troglo-
hole, pit	foramini-, fovea-, putei-	lacco-
hollow	alvei-, cavi-, lacuna-	cystho-
hollow	cavi-	cyto-
hollow	cavi-	cyttaro-
hollow	cavi-	koilo-
hollow	excavati-, -cavus	coelo-
hollow, a	cavi-	gyalo-
hollowed	alvei-	scapho-
hollowed	excavati-	glaphyro-
honey	melli-	meli-
honeycomb	favi-	cerio-
hood	cuculli-	(no Gr. equiv.)
hoof	unguli-	hople-
hook	hami-, unci-	harpago-
hooked	adunci-	grypo-
horn	corni-, cornu-	kero-
horn	-cornu	-ceras
horn	cornu-, cornui-	cero-
horned	cornuti-	cerato-
horny reptile-scale	cornei-	pholido-, -pholis
horse	equi-	hippo-
hot	calidi-	thermo-
hour	hora-	horo-
house	aedi-	-oecium (-oikion)
huge, monstrous, prodigious, terrible	immani-, ingenti-, monstrosi-	peloro-
hummock	cumuli-, grumuli-, tumuli-	soro-
hump, protuberance	gibberi-, gibbi-	hybo-
humped, hunched	gibbi-	cypho-
hurricane	procella-, turbini-	lailapo-
husk	gluma-	lemmato-
husk, rind, shell	cortici-, crusta-, folliculi-	celypho-
husk, rind, shell	gluma-, putamini-, testa-	celypho-
ice	glaciei-	cryo-, crymo-
ice	glaciei-	crystallo-

identical	(idem)	tauto-
immoveable	immobili-	akineto-
in proportion to	proportioni-	symmetro-
increasing	aucti-	aexi-
indistinct	obscuri-	amydro-
inflexible, unbending	rigidi-	atropo-
injury	laesioni-, vulneri-	blabe-
inner	interiori-	esotero-
innermost	intimi-	esotato-
insect	insecti-	entomo-
insignificant	exigui-	asemanto-
interwoven	intexti-	symploco-
iron	ferri-	sidero-
island	insuli-	neso-
islet	(insula parva)	nesido-
isolation	solitudini-	apomono-
jagged	serrati-	carcharo-
jar	dolii-	pitho-
jar	dolio-	amphoreo-
jar	olla-	stamno-
jaundice	icteri-	ictero-
javelin	pili-	akonto-
jaw	maxilli-	gnatho-
jawbone	maxilli-	siagono-
joined together	iuncti-	synapsi-, synapto-
joint	arti-, articuli-	arthro-
joint	commissuri-	armo-
jointed	articulati-	enarthro-
jug, pitcher	urcei-, urceoli-	cyrillio-
juice, moisture, sap	succi-	chylo-, chymo-
juice, sap	succi-	opo-
juice, sap	succidi-, sucidi-	cecido-
junction, union	coniuncti-	arma-, arme-
keel	carina-	tropi-, tropido-
keel	carini-	steira-
keen, pungent, sharp	acri-	drimy-
key	clavi-	cleido-
kid	haedi-, hoedi-	eripho-
kidney	reni-	nephro-
kiln	fornaci-	camino-
kind, of a	-genus	-geneus
king	regi-	basileo-
knee	genu-	gony-
-kneed	geniculati-	gony-
knobbly	nodosi-, torulosi-	gongylo-
knuckle	articuli-	condylo-
ladder	scala-	apobathro-
ladder	scala-	climaco-, -climax
ladle	trulli-	arytero-, arytaino-

85

laid back	supini-	hyptio-
lamb	agni-	amno-, arno-
lamp, light	lumini-	lychni-, lychno-
lamprey	lampetro-, muraeni-	bdello-, -bdella
lance, spear	lancea-	loncho-
land	telluri-, terrestri-	geo-
large	grandi-, grossi-	megalo-
lark	alauda-	corydo-
last	postremi-	opistato-
last	postremi-, ultimi-	prumno-, prymno-
last	postremi-, ultimi-	pymato-
last	ultimi-	eschato-
last year's	annotini-	perysino-
late	serotini-	hystero-
late	serotini-	opse-, opsi-
lattice	cancelli-	cinclido-
lead	plumbi-	molybdo-
leaden	lividi-	pelio-, pelidno-
leaf	folii-	phyllo-
least	minimi-	microtato-
leather	aluta-, cori-, pelli-	byrso-
leather	aluta-, cori-, pelli-	scyto-
leather	aluti-, corii-, pelli-	diphthero-
leathern sack	culei-	coryco-
-leaved	-folia	-phylla
leaven, yeast	fermenti-	zymo-
leech	hirudini-	bdello-, -bdella
leek-green	porracei-	prasino-
left	sinistri-	aristero-
lefthand	scaevi-	scaio-
lefthand	sinistri-	euonymo-
leg	cruri-	scelo-
legging	ocrei-	cnemido-, -cnemis
lemon-yellow	citrini-	citrino-
lentil	lenti-	phaco-, phako-
less	minori-	elasso-, elatto-
less, smaller	minori-	meio-
lesser	minori-	microtero-
letter, writing	litteri-, scripti-	grammato-, -gramma
life	vita-	zoe-
light	luci-	auge-
light	luci-	phengo-
light	luci-	photo-
lightning	fulguri-	astrapo-
lightweight	levi-	elaphro-
like	simili-	-oides
likeness	-aceous, -ago, -eus	-ites, -itis
likeness	imagini-	eikono-
lily	lili-	lirio-, leirio-
limb	arto-, membri-	gyio-
limb	artu-, membri-	colo-
lime	calci-	titano-

limpet	lepadi-	lepado-
line	linei-	-gramme
lion	leoni-	leonto-
lip	labii-	cheilo-
-lipped	-labiatus	-chilus
little ear	auriculi-	otio-
little hole	foramini-	trymalio-, trymatio-
little, small	exigui-, minuti-	baeo-
little, small	parvi-, pusilli-	baeo-
little, small	parvi-	pauro-
liver	iecori-	hepato-
living	-vivus	-bius
lizard	lacerti-	sauro-
lobe	lobi-	lobo-
lobster	astaci-	astaco-
lobster	cammari-, gammari-	cammaro-
long	longi-	dolicho-
long	longi-	macro-
looped	arcuati-	brochido-
loose	laxi-	chalaro-
loosing	solventi-	lysi-
louse	pediculi-	phtheiro-
lovely	amabili-, amoeni-, venusti-	erato-
loving	amanti-	philo-, -philus
low	humili-	chthamalo-
low, poor	demissi-, humili-	tapeino-, tapino-
low, short, small, tiny	brevi-, demissi-,exigui-	elachy-
low, short, small, tiny	humili-, minuti-	elachy-
low, short, small, tiny	parvi-, pusilli-	elachy-
low-growing, on the ground	demissi-, humili-	chamae-, chamelo-
lung	pulmoni-	pneumono-
lying forwards	proni-	prene-
lynx	lynci-	lynco-
lyre	lyra-	lyro-
made	-factus	-poietos
maiden	virgini-	partheno-
maimed	manci-	pero-
mainland	continenti-	epeiro-
making	-ficiens	-poiea, -poiesis, -poios
male	mari-, masculi-	arrheno-, arseno-
man	homini-, viri-	andro-, anthropo-
man	homini-, viri-	broto-
mantle	pallii-	himatio-
many	multi-	poly-
many-coloured	variegati-	poecilo-, poikilo-
marble	marmori-	marmaro-
mark	-nota	-stigma
mark	nota-	charagmato-
mark	signi-	-sema
marked	notati-	characto-
marked	notati-	stigmato-

marrow	medulla-	myelo-
marsh	palustri	heleo-, helo-
marsh, swamp	paludi-	telmato-
marvel, wonder	mirabili-, miri-, mirifici-	thaumato-, -thauma
marvellous, wondrous	mirabili-, miri-, mirifici-	thaumasto-
mat	storei-, tegeti-	phormo-
mattock	bipalii-	marro-
mattress	culciti-	stibado-
maze	ambagi-, labyrinthi-	labyrintho-
meadow	prati-	-leimon, -limon
measure	mensura-	metre-, metrio-, metro-
mellow	miti-	pepono-
membrane	membrana-	-chorion
membrane	membrani-	meningo-
messenger	nuntii-	angelo-
middle	medi-	meso-
mild, soft	leni-, miti-, molli-	prau-
mildew	robigini-	erysibo-
milk	lacti-	galacto-
milking	mulcti-, mulgi-, mulsi-	bdallo-
mill	mola-	mylo-
minute, tiny	exigui-, minuti-	tynno-
minute, tiny	parvi-, pusilli-	tynno-
mirror	speculi-	enoptro-
mirror	speculi-	esoptro-
mixed	mixti-	mixo-
moist	humidi-, made-	hygro-
moisture	humidi-, rori-	icmado-
mole	talpi-	aspalaco-
mole	talpi-	scalopo-
mole	talpi-	spalaco-
mollusc, shell	concha-	concho-
monkey	simia-, simii-	cercopo-
monkey	simii-	pitheco-
moon	luna-	meni-
moon	luna-	seleni-
more	pluri-	pleio-
morning	matutini-	proio-
moss	musc-	bryo-
most, very many	plurimi-	pleisto-
moth	tinei-	phalaino-
mother	matri-	metro-
mound	cumuli-, grumuli-, tumuli-	soro-
mound	tumuli-	tymbo-
mountain	monti-	oreo-, ores-, oro-
mountain-stream	torrenti-	charadro-
mouse	muri-	myo-, -mys
moustache	(no Latin equivalent)	mystaco-
mouth	ori-	stomato-, -stoma
much	multi-	pletho-
mucus	pituita-	phlegmato-
mud	limi-	ily-, ilyo-

murder	neca-, -cidus	phono-
murder	neci-, -cidus	ctono-, -ctonos
muscle	musculi-	myo-, -mys
muscle	tori-	ino-
mussel	mitylli-	myako-
mutual	mutui-	allelo-
myrrh	murra-	smyrno-
nail	clavi-	gompho-
name	nomini-	onomato-
narrow	angusti-	lagaro-
narrow	angusti-	steno-
navel	umbilici-	omphalo-
near	propinqui-	anchi-
near, close by	iuxta-	pros-
near, neighbour	proximi-, vicini-	plesio-
near, nigh	comini-, propinqui-	schedo-
neat	eleganti-	compso-
neck	cervici-	isthmo-
neck	cervici-, colli-	trachelo-
necklace	monili-	hormo-
needle	acu-	-belone
needle	acu-	rhaphido-
neighbour	vicini-	-geiton, -geton, -giton
net	cassi-	arcy-
net	reti-	dictyo-
new	novi-	caeno-, ceno-
new	novi-, novo-	neo-
next to	iuxta-	meta-
night	nocti-	nycti-, nycto-
nipple	papilli-	thele-
nipple	papilli-	tittho-
nodding	nutanti-, nutati-	brizo-
noisy	strepiti-	psopho-
noisy	strepitu-	spharago-
nook	recessi-	mycho-
noon	meridiei-	mesembria-
noose	laquei-	brocho-
north	septentrionali-	boreo-
northwest	cauri-, cori-	zephyro-
nose	nasi-	myctero-
nose	nasi-	rhino-
nourished	aliti-	threpsi-, threpto-
now	(iam, nunc)	arti-
nut	nuci-	caryo-, -caryon
oar	remi-	cope-
oats	aveni-	aegilopi-
oblong shield	scuti-	hoplo-
oblong shield	scuti-	thyreo-
ochreous, pale yellow	ochracei-, sili-	ochro-
odd	impari-	perisso-

oil	olei-	elaio-
oil	olei-	lipa-
ointment	collyrii-, unguenti-	myro-
old	seni-, veteri-	geraio-, geronto-
old	veteri-, vetus-	palaeo-
old sow	scrofa-	gromphado-
old, ancient	veteri-	archaeo-, archaio-
on both sides	ambi-	amphi-
once	singuli-	hapax-
one-&-a-half	sesqui-	triemi-
opening	ori-	-poros
opposite	adversi-, contra-	enantio-
orange-red	rufi-	pyrrho-
order	ordini-	phylo-
organization	ordinationi-	systemato-
ostrich	struthioni-	strutho-
other	alii-	allo-
outermost	extremi-	pymato-
outside	extra-	ecto-, exo-
oven	furni-	ipno-
owl	strigi-	strigo-
ox	bovi	boö-, bou-, bu-
paired	binati-	dicha-, dicho-
paper	charta-	charto-
paper	charta-	papyro-
parasol	umbelli-	sciado-
parrot	psittaci-	psittacin-
part	parti-	-meros
partridge	perdici-	perdico-
pasture	pascui-	nomo-, -noma
path, track	calli-, semita-, tramiti-	celeutho-
peacock	pavoni-	tao-
pearl	margarita-	margarito-
pebble	calculi-	cachleco-
pebble	calculi-	chalico-
pebble	calculi-	psepho-
pedigree	stemmati-	genea-
peel, rind	cuti-, tunica-	lepismato-, lepyro-
peel, rind	cuti-, tunica-	lopo-
peg, rod	feruli-, paxilli-, virga-	cercido-
perforated	foraminati-, perforati-	tremato-, treto-
perforated	perforati-	ethmoideo-
perforation	foramini-	tresi-
perfume	odori-	myro-
pickaxe	dolabri-	orygo-
pig	porci-	choero-, choiro-
piglet	porcelli-	choiridio-
pillar	columni-	ciono-
pillar	columni-	stelo-, -stele
pillar, pole	columni-	stylo-
pimple, tubercle	pustuli-, tuberculi-	chalazo-

pin	acu-	perono-, -perone
pine-cone, top	coni-, strobili-	strobilo-
pipe	tubi-	ocheto-
pipe	tubi-	siphono-, -siphon
pipe	tubi-	soleno-, -solen
pipe	tubi-	syringo-, -syrinx
pit	fovei-	bothrio-
pit	fovei-, putei-	barathro-
pit	siri-	siro-
pitch	pici-	pisso-
pivot	cardini-	cnodaco-
pivot	cardini-	polo-
place	loci-	topo-
place, plot, region, space	regioni-	choro-
plain, simple	simplici-	aphelo-
plaited, woven	intexti-, plicato-	pleco-, plecto-
plaiting, twisting, weaving	torti-	plegmato-, -plegma, plexi-
plant	planti-	phyto-
plate	patelli-	placo-
plate, sheet	bractei-, catilli-	elasmo-
plate, sheet	lamini-, patelli-	elasmo-
plaything	ludibri-	-paegma
pleasant-scented	fragranti-	hedyosmo-, hedypnoi-
plough	aratri-, vomeri-	arotro-
ploughshare	vomeri-	hyno-
pod	siliqua-	celypho-
point	aculei-	-glochin
point (dim.)	aculeoli-	glochidio-
point	acumini-	centro-, -centron
point	acumini-, cuspidi-	aci-, acido-
point	mucroni-, puncti-	aci-, acido-
point	acumini-, cuspidi-	aechmo-, aichmo-
point	mucroni-, puncti-	aechmo-, aichmo-
point, spike	cuspidi-	stortho-, storthyngo-, storyne-
point, summit, tip	acumini-, apici-	-acme
point, summit, tip	cacumini-, summi-	-acme
pointed	acuti-	cestroto-
pointer	indici-	gnomono-
poison	veneni-	ios-
poison	veneni-	pharmaco-, pharmako-
polished, smooth	levi-, politi-	xesto-
pond	stagnali-	limno-
pond, pool	lacu-, stagni-	telmato-
pond, pool	stagni-	tipho-
porcupine	hystrici-	hystricho-
pot	olla-	chytro-
pottery	fictili-, figlini-	chytro-
pouch	marsupi-, sacco-	pera-
press	pressi-	thlipso-
pricked	puncti-, pungenti-	stizo-
prodigious	immani-	perisso-

prodigy	monstrosi-	terato-
prop, stay, support	adminiculi-, columini-	ereismato-
prop, stay, support	firmamenti-, fulcimenti-	ereismato-
prop, stay, support	pedamenti-	ereismato-
prop, stalk, support	-firmamentum	-pelma
prop, support	fulti-	sterigmo-
protuberance	umboni-	exoncomato-, exoncosi-
pruning	putationi-	cladeuto-
pulse	legumini-	osprio-
pup	catelli-	scylaco-
purple	murici-	calche-
purple	purpureo-	porphyreo-, porphyro-
quail	coturnici-	ortygo-
quick, rapid, swift	celeri-, citi-, rapidi-	tachy-
quiet	taciti-	hesychio-
quince-yellow	lutei-	melino-
quiver	pharetra-	goryto-
quiver	tremuli-	einosi-
quoit	tori-	disco-
race	geni-	phylo-
ragged	laciniati-	trychero-, trychino-
ragged, torn	lacerati-	rhacoi-, -rhacodes
rain	pluvii-	hyeto-, -hyet
rain, shower	imbri-	ombro-
raised	elevati-	hyperphoro-
raised	prominenti-	execho-
raisin	astaphi-	staphido-
ram	arieti-	crio-
rank, row	-farius, ordini-, serie-	sticho-, -stichos
rank, row	-farius, ordini-, serie-	stoechado-, stoecho-
rattle	crepitu-	crotalo-
raven	corvi-	coraco-
ray	radii-	actino-
razor	novaculi-	xyro-
red	rubri-	erythro-
reed-plume	paniculi-	antheli-
remarkable	insigni-, praecipui-	perisso-
removeable covering	lanugini-	-achne
reproductive organs	genitali-	-gone
resin	resina-, resini-	rhetino-
returning, returned	reverti-	anacamps-, anacampt-
rib	costa-	pleuro-
ridge	jugi-	lopho-
righthand	dextra-	dexio-
rind	crusta-	scirrho-
ring	annuli-	cyclo-
ring	annuli-, circuli-	crico-
ripe	maturi-	horio-
ripe	maturi-	pepeiro-
rising	assurgenti-	anabaeno-

river	amni-, flumini-, fluvii-	potamo-
road, way	cursu-, itineri-, via-	celeutho-
rock	rupi-, saxi-	petro-
rock	scopuli-	scopelo-
rod	virga-	canono-
rod	bacilli-, virga-	rhabdo-
rod	bacilli-, virga-	rhapido-
rolled up	involuti-	aneilemato-
roller	cylindri-, scutuli-	cylindro-
roof	tecti-	orophe-
roofing	teguli-	erepsi-
root	radici-	rhizo-, -rrhiza
rope	funi-, resti-	schoeno-
rope	funi-, resti-	-sira, -seira
rose	roseo-, rosi-	rhodo-
rotten	putridi-	sapro-
rotten	putridi-	sathro-
rough	asperi-	cerchno-
rough, rugged	asperi-, horridi-	trachy-
rough, scabby	scabri-	psoro-
round	circulari-, rotundati-	gyro-
round	globosi-, glomerosi-	gongylo-
round	orbiculati-, rotundi-	gongylo-
round	orbiculati-	strongylo-
round shield	clipea-	aspido-, -aspis
rubbed, worn	fricti-, triti-	trycho-, tryo-, tryso-
rubbed, worn	triti-	tribaco-, tribo-
rudder	gubernaculi-	pedalio-
rump	cluni-	glouto-
rump	cluni-	pygo-
runaway, a	fugitivi-	drapeto-
running	-currens	-dromos
S-shaped	(no Latin equivalent)	sigmato-, sigmoideo-
sabre	acinaci-	machairo-
sacred	sacri-, sancti-	hiero-
saddle	sella-	sagmato-
saffron-coloured	crocei-	crocoto-
sail	carbasi-, veli-	histio-, histo-
salt	sali-, salso-	halo-
same	(idem)	homo-
sand	arena-, sabuli-	psammo-
sand	arena-, sabuli-	thino-
sand	areni-	ammo-
sandal, shoe	calce-	pedilo-
sandal, slipper	calceoli-, crepidi-, solei-	crepido-
sauce	condimenti-	zomo-
sausage	botuli-	allanto-
saw	serri-	priono-
scanty	rari-	spanisto-
scar	cicatrici-	oteilo-
scarlet	coccinei-	coccino-

scattered	sparsi-	scedasto-
scattered	sparsi-	sporadico-
scion, slip	surculi-	oscho-, -osche
scorpion	scorpii-	scorpio-
scraped	rasi-	xysto-, xystro-
scurf	scabie-	leicheno-
sea	mari-	ponto-
sea	mari-	thalasso-
sea, of the	marini-	enalio-
sea, of the	marini-	pelago-, pelagio-
sea-monster	ceti-	ceto-
seal, stamp	sigilli-, signi-	sphragido-
sealed, stamped	sigillati-, signati-	sphragisto-
seam	suturi-	rhaphi-
seaside, by the	maritimi-	paralia-
seasonable	hornotini-, tempestivi-	horaio-
seat	sedi-, sedili-	-hedra
seated	sessili-	hedraio-
seaweed	algi-	phyco-
second	secundi-	deutero-
secret	abditi-, arcani-, conditi-	lathro-
secret	occulti-, remoti-, secreti-, tecti-	lathro-
secret	abditi-, arcani-, occulti-	crypsi-, crypso-
secret	remoti-, secreti-	crypsi-, crypso-
seed	semini-	spermato-
seed	semini-	sporo-
-seeded	-semina	-sperma
seen together	conspectu-	synopsi-
seizing	prehensi-	harpazo-
self	ipsi-	auto-
sewn	suti-	rhammato-
sewn, stitched	suti-	rhapso-, rhapto-
shade, shadow	umbra-	scia-, scio-
shaded	umbrati-	scioto-, -sciodes
shaft, stem	scapi-	scapo-
shaggy	hirsuti-, hirti-, hispidi-, villi-	dasy-
shaggy	hirsuti-, hirti-, villi-	trachy-
shaggy	villosi-	lasio-
shallows, shoal	vadi-	tenago-
shape	formi-, -formis	morpho-, -morpha
sharp	acri-	oxy-
sharp point	acuti-	stonycho-
she-goat	capra-	chimaero-
sheaf, truss	fasci-, manipuli-, mergiti-	-dragma
shears	forfici-	psalido-
sheath	vagini-	coleo-
sheep	ovi-	melo-
shell	concha-	concho-
shell	crusta-	ostraco-
shelter	tecti-	-scepasma
shelter	tecti-	stegano-, -stegia, stego-
shield-boss	umboni-	exoncomato-, exoncosi-

shield, oblong	scuti-	hoplo-, thyreo-
shield, round	clipea-	aspido-, -aspis
shin	tibii-	cneme-
shining	nitenti-	phaedro-, phaeno-
shining	nitenti-	phano-
shiny black	nigri-	melancho-, melano-
shoot	surculi-	blasto-, -blastos
shoot	surculi-	clono-
shoot	surculi-	ormeno-
shoot	surculi-	oscho-, -osche
shoot	virga-	rhadamno-
shoot, sucker	surculi-	ptortho-
short	brevi-	brachy-
short	brevi-	conto-
shoulder	humeri-	omo-
shoulder-blade	scapuli-	omoplato-
shovel	pala-	scapano-, -scapane
shrew	sorici-	hyraco-
shrivelled	rugosi-	rhikno-
shrub	frutici-	thamno-
sickle	falci-	drepano-
sickle	falci-	zangklo-
side	lateri-	pleuro-
sieve	cribelli-, cribri-	coscino-
silk	sericei-	serico-
silkworm	bombyci-	bombyco-
silver	argenti-	argyro-
similar	simili-	homoio-
simple	simplici-	lito-
single	soli-	haplo-
single	uni-	mono-
skin	cuti-, pelli-, -pellis	dermato-, dermo-, -derma
slab, tablet	tabuli-	placo-
slack	laxi-	laparo-
slack	laxi-, remissi-	aneto-, aneuro-
slanting	obliqui-	loxo-
slanting	obliqui-	plagio-
slender	gracili-	ischno-
slime	muci-	borboro-
slime	muci-	myxo-
slough off	exuti-, exuvi-	ecdysi-
small	parvi-	micro-
small	parvi-	tyttho-
small cup	poculi-	cymbi-
small hair	puberuli-	trichio-
small round shield	clipeo-	pelto-
small tongue	linguli-	glossario-, glossidio-
smallest	minimi-	elachisto-
smell	odori-, olor-	osm-, osmo-, -osme
smoke	fumi-	capno-
smoke	fumi-	typho-
smooth	laevi-, laevigati-	leio-

smooth	laevi-, laevigati-	lispo-, lisso-
snail	cochlea-	cochlio-
snail	cochlei-, conchi-	strombo-
snake	angui-, serpenti-	herpeto-
snake	serpenti-	ophio-
sneeze	sternuti-	ptairo-, ptaero-
snow	nivi-	chiono-
snow	nivi-	niphado-
snub-nosed	simi-	simo-
soap	saponi-	smegmato-
soft	leni-, miti-, molli-	malaco-
soft	molli-	hapalo-
soft	molli-	tryphero-
soil	soli-	pedo-
sole	planti-	tarso-
sole of the foot	planta-, planti-	pelmato-
soot	fuligini-	lignyo-
sound	soni-	phono-, -phone
south	australi-, meridionali-	noto-
sow	sui-	hyo-, syo-
span	dodranti-	spithame-
spear	hasti-	aechmo-, aichmo-
spear	hasti-	akonto-
spear	hasti-	dory-
spice	aromati-	aromato-
spider	aranei-	arachno-
spindle	fusi-	atracto-
spindle	fusi-	clostero-
spine	spini-	acantho-
spiral	cochlei-	helico-
spirit	anima-	pneumato-
spit	veru-	obelo-
spitting, a	spuenti-	ptysi-
spittle	sputi-	ptysmato-
spleen	lieni-	spleno-
splendour	magnifici-	aglaio-, aglao-
splinter	assuli-, scinduli-	schidio-
split	fissi-, scissi-	schis-, schismato-, schisto-
split	fissi-, scissi-	schizo-
sponge	fungi-	sompho-
sponge	fungi-	spongio-, spongo-
sponger	parasiti-	colaco-
spoon	cochleari-	mystro-
spot	gutta-, macula-	balio-
spot, stain	macula-	spilo-
spotless	casti-, immaculati-, puri-	aspilo-, aspiloto-
spotted	guttati-, maculati-, punctati-	sticto-, -sticta
spotted	guttati-, maculati-, punctati-	stizo-
Spring	veri-, verni-, vernali-	earo-
spring, a	fonti-, scaturigini-	pego-
spur	calcari-	centro-, -centron
squid	loligini-	teuthido-

squirrel	sciuri-	sciuro-
stable, stall, shelter	stabuli-, tegmini-	stathmo-
staff	baculi-, fusti-, scipioni-	sceptro-
stag-beetle	scarabaei-	carabo-
stain	macula-	celido-
stake	pali-	scolo-, scolopo-
stake	pali-	stalico-, stalido-
stake	-sudes, -vallus	characo-, -charax
stamen, thread	stamini-, -stamineus	stemono-, -stemon
standard	vexilli-	semeio-, -semeia
star	astri-, stelli-	phostero-
star	stelli-	astro-
starch	amyli-	amylo-
-starred,starry	stellati-	astero-
stem	cauli-, -caulis	caulo-, -caulon
stem	scapi-	scapo-
stem, stump, block	stipiti-, stirpi-	stypo-
step, stair	gradi-, scali-	bathmo-
stick, wood	ligni-	phrygano-
sticky	visci-	ixod-
stilt	gralli-	colobathro-
sting	stimulanti-, urenti-	dako-, -dakos
stinking	foetidi-	bromo-
stinking	foetidi-, graveolenti-	ozo-
stitched	suti-	cesto-
stock	caudici-	pythmeno-
stock	caudici-	stelecho-
stone	lapidi-, saxi-	litho-
stork	ciconii-	pelargo-
storm	procella-	thyello-
stout, strong	robusti-, validi-	alkimo-
straight	recti-	euthy-
straight	recti-	ithy-
straight	recti-	ortho-
straight sword	ensi-	xiphe-
strangle	stranguli-	-anche
strap, thong	lori-	himanto-
straw	straminei-	carpho-
stream	flumini-	rheithro-
stream	torrenti-	rhyaco-
strength	robusti-, validi-, viri-	stheno-
stretched	extensi-, extenti-	ecteino-, ecteno-
stretched	tensi-, tenti-	teino-, tino-
stretched out	extenti-	tany-
stretched out	extenti-	tetano-
strike	inflicti-	plectro-
striped	vittati-	rhabdoto-
strong, stout	robusti-, validi-	alkimo-
strong	robusti-, validi-	cratero-, crato-
strong	validi-	iphi-
strong	validi-	ischyro-
strong	validi-	sterrho-

stuffed	farcti-	sacto-
stump, trunk	caudici-, stipiti-	premno-
suet, tallow	sebi-	stearo-, steato-
suffering	passi-	patho-
summer	aestati-	therei-, thero-
sun	soli-	helio-
swallow	hirundini-	chelidono-
swan	cygni-	cycno-
swathed	incunabuli-	spargano-
sweep	versi-	saro-
sweet	dulci-	glyco-
swift	celeri-, citi-, rapidi-, veloci-	ocy-
swimming	natanti-	colymbi-
swimming	natanti-	necto-
swollen	tumidi-	cymato-
swollen	tumidi-	oedo-
table	mensa-	trapezo-
tablet	tabula-	schede-
tablet	tabuli-	pinaco-, -pinax
tail	cauda-	cerco-
tail, -tailed	caudi-, caudati-	uro-, -ura, -oura
tapering	fastigiati-	phoxi-
target	parma-	pelto-
tawny	fulvi-	cirrho-, cirro-
tawny, yellow-brown	fulvi-	xutho-
tear	lacrimi-	dakryo-
tendril	capreoli-, clavicula-	helino-
testicle	testi-	orchi-
thick	crassi-, spissi-	pachy-
thicket	dumeti-	lochmaio-
thicket	dumeti-	tarpho-
thigh	femori-	mero-
thin	rari-	araio-
thin	tenui-	lagaro-
thin	tenui-	lepto-
thin-skinned	membrani-	hymeno-
thorn	senti-	akaino-
thread	fili-	mito-
thread	fili-	nemato-
thread	fili-	rhammato-
three-cornered	triquetri-	trigono-
threshold	limini-	bathmo-
throat	fauci-	laimo-
throat	fauci-	laryngo-
throat	fauci-	pharyngo-
through	per-	dia-
throw	iactu-	bolo-, -bolos
throw	iactu-	rhipso-
throw out	eiecti-	ecballo-
thrown	iacti-	rhipto-
thunder	tonitru-	bronto-

thunderbolt	fulmini-	cerauno-
tiger	tigri-	tigro-
time	tempori-	chrono-
tin	stanni-	cassitero-
toad	bufoni-	phryno-
together	co-, simul-	syll-, sym-, syn-, syrrh-, sys-, syz-
together with	co-, col-, com-, con-, cor-	hama-
tongue	lingui-	glosso-, glotto-
tooth	denti-	odonto-
top	apici-	colopho-
top	culmini-, summi-	acro-
top	culmini-, summi-	corymbo-
top	culmini-, summi-	corypho-
top	turbini-	rhombo-
top	turbini-	strombo-
torch	faci-, funali-, taeda-	lampado-
torn	lacerati-	drypto-
torn	lacerati-	sparasso-, sparatto-
torn, mangled, rent	lacerati-	amyxi-
tortoise	testudini-	chelono-
tower	turri-	pyrgo-
tower	turri-	tyrsio-
track	calli-	ichno-
trap, part of a	tendicula-	schastero-
tree	arbori-	dendro-
treetrunk	stirpi-	cormo-
trellis	cancelli-	cinclido-
triangular	triangulari-	delto-
tribe	tribu-	phylo-
trident	tridenti-	thrinaco-, -thrinax
trough	alvei-	leno-
truffle	tuberi-	hydno-
truth	veri-	aletho-
tunic	tunica-	-chiton
turn	flexi-, torquei-, versi-	trepo-
turned	flecti-, torti-, verti-	trepsi-, trepto-
twig	surculi-	oscho-, -osche
twig	virga-	canono-
twin	bini-, gemini-	didymo-
twisted	cochleati, torti-	helicto-
twisted	torti-	streblo-, stremmato-
twisted	torti-	strepho-, strepsi-, strepto-
twisted	torti-	stropho-
unarmed	inermi-	anoplo-
unborn, unformed	foeti-, inchoati-	embryo-
uncommon	rari-	perisso-
undignified	ignobili-	asemno-
unequal	inaequi-	aniso-
unequal	inaequi-	scaleno-
uneven	inaequi-	perisso-

uneven	inaequi-, iniqui-	anomalo-
united	connati-	gamo-
unknown	ignoti-	adelo-
unknown	ignoti-	agnoto-, agnosto-
unmarried	caelibi-	agamo-
unripe	immaturi-	omphaco-
unseen	invisi-	aphano-
untilled	inculti-	atropo-
unusual	insueti-	aethio-, -eo-
up	sub-	ana-, ano-
upon	super-	epi-
uppermost	summi-	hypato-
variegated	maculosi-	eusticto-
vault	fornici-	camaro-
veil	veli-	calyptro-
vein	vena-	phlebo-, -phleps
vermilion	minii-	milto-
very	per-	za-
very big	maximi-	megisto-
vessel	vasi-	angio-, ango-
vine-prop	pedamini-, pedamenti-	camaco-
violet	viola-	io-
violet-coloured	violacei-	ianthino-
wall	muri-, parieti-	teicho-, toicho-
wand	vimini-, virga-	thyrso-
wart	verruci-	phymato-
wasp	vespa-	spheco-
water	aqua-	hydato-, hydro-
water-repellent	(no Latin equivalent)	adianto-
wax	cera-	cero-
way	via-	hodo-
weasel	musteli-	gale-, gali-
web, woven	textili-	plocio-, ploco-, -ploca, -ploce
wedge	cunei-	spheno-
week	hebdoma-	hebdoma-
weevil	curculioni-	cio-
weevil	curculioni-	trogo-
well	bene-	eu-
west	occidenti-	hespero-
wheat	tritici-	pyro-
wheel	roti-	trocho-
whip	flagelli-	mastigo-
whispering	susurri-	psithyro-
white	albi-	argi-, argo-
white	albi-	leuco-
whole	cuncti-, toti-	ulo-
wick	fili-	thryallido-
wild	sylvestri-	agrio-
wild beast	feri-	thero-, therio-

wild boar	apri-	capro-
willow-twig	vimini-	lygo-
wind	venti-	anemo-
wind	venti-	pneumato-
wine	vini-	oeno-
winecup	poculi-	phiali-, phialo-
winecup	poculi-	poterio-
wing	ala-	pterygo-, -pteryx
winged	alati-	ptero-
winnowing-fan	vanni-	likno-
winnowing-fan, -shovel	vanni-	ptyo-
winter	hiemi-	cheimo-
wintercherry	physali-	halicacabo-
within	intra-	endo-, ento-
without	(sine)	lipo-
without	e-, ex-	a-, an-
wolf	lupi-	lyco-
womb	uteri-	coeleo-, coelia-, coelio-
womb	uteri-	delphy-, delphyo-
womb	uteri-	hystera-
womb	uteri-	koilia-, koilio-
wood	ligni-	xylo-, -xylon
woodworm	cossi-	thripo-, -thrips
wool	lani-	eiro-
wool	lani-, velleri-	mallo-
woollen	floccosi-, lanati-	erio-
world, universe	mundi-	cosmo-
worm	vermi-	helmintho-
worth	digni-	axi-, axio-
wretched	miserabili-	talaeporo-
wrinkle	rugi-	rhyti-, rhytido-
wrinkled	rugosi-	rhyso-, rhysso-
Y-shaped	(no Latin equivalent)	hypsilo-
yearly	annui-	epeteio-, eteio-
yoke	jugi-, -jugus	zygo-
yoke-attachment	jugi-	zeuglo-
yoked	jugali-, jugati-	zeugitido-
yoked	jugali-, jugati-	zeukto-
yoking	jungenti-	zeuxi-

PLANT-NAME SUPPLEMENT

GREEK	LATIN	ENGLISH
achras	pyrus	pear
aigeiros	populus	poplar
aigilops	avena	oat
aira	lolium	darnel
aischunomene	mimosa	wattle
akanthion	onopordon	cottonthistle
aktea	sambucus	elder
alkea	malva	muskmallow
althaia	althaea	marshmallow
amarantos	amaranthus	love-lies-bleeding
ampelos	vitis	vine
amygdalea	pr.amygdalus	almond
anacampseros	sedum	stonecrop
andrachne	peplis	purslane
androsaimon	androsaemum	tutsan
anthyllis	cressa	kidneyvetch
aparine	galium	cleavers
apion, apios	pyrus	pear
argemone	papaver	poppy
aria	qu.ilex	holm-oak
arisaron	arum	cuckoopint
arkeuthos	juniperus	juniper
asaron	asarum	asarabacca
askyron	hypericum	St. John's wort
aspalathos	astragalus	tragacanth
asparagos	asparagus	asparagus
asphodelos	asphodelus	kingspear
astragalos	astragalus	tragacanth
atraktylis	carthamus	safflower
atraphaxis	atriplex	orache
bakkaris	conyza	fleabane
ballote	ballota	horehound
batos	rubus	bramble
batrachion	ranunculus	buttercup
bliton	amaranthus	love-lies-bleeding
borassos	borassus	fanpalm
brabylon	prunus	sloe
brabylon	prunus	damson
briza	secale	rye
bromos	avena	oats
byblos	papyrus	papersedge
chairephyllon	chaerophyllum	chervil
chamaiakte	s. ebulus	dwarf elder
chamaibalanos	e. apios	acorn-spurge
chamaidaphne	ruscus	dwarf laurel
chamaidrys	teucrium	germander

chamaigeiron	tussilago	coltsfoot
chamaikerasos	p. cerasus	dwarf cherry
chamaikissos	glechoma	groundivy
chamaikyparissos	chamaecyparis	groundcypress
chamaileuke	tussilago	coltsfoot
chamaimelon	chamaemelum	chamomile
chamaimyrsine	ruscus	butcher's broom
chamaipeuke	chamaepeuce	groundlarch
chamaipitys	ajuga	groundpine
chamaisyke	chamaesyce	groundspurge
chamaizelon	genista	greenweed
chamelaea	d. oleoides	dwarf-olive
chiliophyllos	millefolium	milfoil
daphne	laurus	laurel
delphinion	delphinium	larkspur
dendrolibanos	rosmarinus	rosemary
dendromalache	lavatera	treemallow
dipsakos	dipsacus	teazel
donax	arundo	reed
doryknion	convolvulus	bindweed
drabe	draba	whitlowgrass
drys	quercus	oak
ebenos	ebenus	ebony
echinopous	echinops	globethistle
echion	echium	bugloss
elaia	olea	olive
elate	abies, picea, pinus	fir, spruce, pine
elatine	kickxia	toadflax
empetron	saxifraga	saxifrage
ereike	erica	heather
ereuthedanon	rubia	madder
erigeron	senecio	groundsel
euonymos	euonymus	spindle
gnaphalion	gnaphalium	cudweed
halikakabon	physalis	wintercherry
hedysaron	coronilla	crownvetch
helichrysos	helichrysum	everlasting
helleboros	helleborus	hellebore
helxine	parietaria	pellitory
herpyllon	thymus	thyme
hierakion	hieracium	hawkweed
horminon	horminum	clary
hypoglosson	ruscus	butchersbroom
hyssopos	hyssopus	hyssop
iasione	convolvulus	bindweed
iberis	lepidium	pepperwort
ion	viola	violet

iris	iris	iris
itea	salix	willow
ixos	viscum	mistletoe
kachrys	hordeum	barley
kalaminthos	calamintha	catmint
kanna	bambusa	cane
kannabis	cannabis	hemp
kapparis	capparis	caper
kardamis	cardamine	bittercress
kardamon	cardamine	bittercress
kastanos	castanea	chestnut
kaukalis	caucalis	burparsley
keanothos	cirsium	thistle
kedros	cedrus	cedar
kedrostis	bryonia	bryony
kegchros	milium	millet
kelastra	ligustrum	privet
kentaurie	centaurium	centaury
kentromyrsine	ruscus	butcher's broom
kerasos	pr.cerasus	cherry
kerkis	populus	poplar
kichoreion	cichorium	chicory, succory
kiki	ricinus	castor
kinara	cynara	globe artichoke
kirkaia	circaea	enchanter's nightshade
kissos	hedera	ivy
kisthos	cistus	rockrose
klinopodion	clinopodium	basil
klymenon	convolvulus	bindweed
knaphos	dipsacus	teazel
knekion	origanum	marjoram
knekos	carthamus	safflower
knide	urtica	nettle
kokkos	qu.coccifera	kermes-oak
kolokynthe	cucurbita	pumpkin
komaros	arbutus	strawberry tree
koneion	cicuta	hemlock
konyza	conyza	fleabane
korchoros	anagallis	pimpernel
krambe	brassica	cabbage
kranon	cornus	dogwood
krataigos	crataegus	hawthorn
krinon	lilium	lily
krithe	hordeum	barley
krithmos	crithmum	samphire
krokos	crocus	saffron
kromyon	allium	onion
kroton	ricinus	castor
kyamos	phaseolus	bean
kydonia	cydonia	quince
kyklamis	cyclamen	sowbread

kyminon	cuminum	cummin
kyparissos	cupressus	cypress
kytisos	medicago	medick
lapathon	rumex	sorrel
larix	larix	larch
lathyros	lathyrus	vetchling
leirion	lilium	whitelily
libanos	tus	frankincense
libanotis	rosmarinus	rosemary
linon	linum	flax
lysimachion	lysimachia	loosestrife
malache	malva	mallow
mandragoras	mandragora	mandrake
mekon	papaver	poppy
melampyron	melampyrum	blackwheat
melea	malus	appletree
melon	malus	apple
meon	meum	spignel
mespile	mespilus	medlar
milos	taxus	yew
mnion	mnium	moss
moron	morus	black mulberry
myrike	tamarix	tamarisk
myriophyllon	myriophyllum	watermilfoil
myrrhis	myrrhis	sweet cicely
myrsine	myrtus	myrtle
myrtos	myrtus	myrtle
nardos	nardus	nard
narkissos	narcissus	narcissus
narthex	ferula	giant-fennel
oa, oia, oua	s. torminalis	service-tree
ogchne, onchne	pyrus	pear
okimon	ocimum	basil
olynthos	grossus	winterfig
olyra	frumentum	corn
onobrychis	onobrychis	sainfoin
onopordon	onopordon	cottonthistle
origanon	origanum	marjoram
ornithogalon	ornithogalum	Star of Bethlehem
orobagche	orobanche	broomrape
orobagche	cuscuta	dodder
orobos	orobus	bittervetch
ostrya	ostrya	ironwood
osyris	chenopodium	goosefoot
oxymyrsine	ruscus	butcher's broom
paionia	paeonia	peony
paliouros	paliurus	Christsthorn

paronychia	reduvia	whitlow
peganon	ruta	rue
peperi	piper	pepper
peplis	eu.peplis	purple spurge
persike	persica	peach
petasites	petasites	butterbur
petasites	tussilago	coltsfoot
petroselinon	petroselinum	rockparsley
peuke	picea	spruce
peukedanon	peucedanum	hogsfennel
phaselos	phaseolus	kidneybean
phegos	fagus	beech
phellos	qu.suber	cork-oak
pheos	poterium	burnet
philyra	tilia	linden
phlomos	verbascum	mullein
pisos	pisum	pea
pistake	pistacia	pistachio
pityoussa	eu.pithyusa	balearic spurge
pitys	pinus	pine
platanos	platanus	plane
poa	gramen	grass
prasion	marrubium	horehound
prinos	ilex	holm-oak
prionitis	betonica	betony
proumne	prunus	plum
ptarmike	achillea	yarrow
ptelea	ulmus	elm
pyrethron	parietaria	pellitory
pyros	triticum	wheat
pyxos	buxus	box
rhachos	rosa	briar
rhamnos	rhamnus	buckthorn
rhaphanis	raphanus	radish
rhapys, rhaphys	rapa, rapum	turnip
rhoa	punica	pomegranate
rhodon	rosa	rose
rhoias	papaver	poppy
rhous	rhus	sumach
santalon	santalum	sandalwood
schinos	lentiscus	mastic
schoinos	schoenus	rush
selinon	petroselinum	parsley
seriphos	artemisia	wormwood
seris	endivia	endive
sesame	sesamum	sesame
sesamon	sesamum	sesame
seselis	tordylium	hartwort
sikyos	cucurbita	cucumber
silignion	siligo	winterwheat

silphion	laserpitium	laser
sinapi	sinapis	mustard
sion	sium	waterparsnip
sisymbron	mentha	mint
sitos	triticum	wheat
skamonia	convolvulus	scammony
skandix	anthriscus	chervil
skarphe	helleborus	hellebore
skilla	scilla	squill
skindapsos	hedera	ivy
skolymos	cynara	globe artichoke
skordion	teucrium	germander
skorodon	allium	garlic
smilax	convolvulus	bindweed
smyrnion	smyrnium	alexanders
sogchites	hieracium	hawkweed
sogchos	sonchus	sowthistle
spartion	spartium	med.broom
spartos	genista	broom
sphagnos	muscus	moss
sphakos	salvia	sage
sphalax	frangula	buckthorn
sphendamnos	acer	maple
sphondylion	heracleum	cowparsnip
spondias	insititia	bullace
stachys	frumentum	corn
statike	statice	sealavender
strychnon	withania	sleepy nightshade
strychnos	solanum	nightshade
sukaminos	morus	white mulberry
sykea	ficus	fig
sykomoros	f. sycomorus	fig-mulberry
telis	trigonella	fenugreek
terminthos	terebinthus	turpentine-tree
tetragonia	euonymus	spindle-tree
teukrion	teucrium	germander
thaliktron	thalictrum	meadowrue
thapsos	cotinus	smokebush
thelypteris	thelypteris	ladyfern
thlaspis	capsella	shepherd's-purse
thridakine	lactuca	lettuce
thryon	juncus	rush
thua, thuia	juniperus	juniper
thyaros	lolium	darnel
thymbra	satureia	savory
thymelaea	gnidium	healwood
thymon	thymus	thyme
tithymalos	euphorbia	spurge
tordylon	tordylium	hartwort
xanthion	xanthium	burrweed

Three-language list of botanical name components

zeia	far	spelt
ziggiberis	zingiber	ginger
zizanion	lolium	darnel
zizyphon	zizyphus	jujube

LATIN	ENGLISH	GREEK
abies, picea, pinus	fir, spruce, pine	elate
acer	maple	sphendamnos
achillea	yarrow	ptarmike
ajuga	groundpine	chamaipitys
allium	onion	kromyon
allium	garlic	skorodon
althaea	marshmallow	althaia
amaranthus	love-lies-bleeding	amarantos
amaranthus	love-lies-bleeding	bliton
anagallis	pimpernel	korchoros
androsaemum	tutsan	androsaimon
anthriscus	chervil	skandix
arbutus	strawberry tree	komaros
artemisia	wormwood	seriphos
arum	cuckoopint	arisaron
arundo	reed	donax
asarum	asarabacca	asaron
asparagus	asparagus	asparagos
asphodelus	kingspear	asphodelos
astragalus	tragacanth	aspalathos
astragalus	tragacanth	astragalos
atriplex	orache	atraphaxis
avena	oat	aigilops
avena	oats	bromos
ballota	horehound	ballote
bambusa	cane	kanna
betonica	betony	prionitis
borassus	fanpalm	borassos
brassica	cabbage	krambe
bryonia	bryony	kedrostis
buxus	box	pyxos
calamintha	catmint	kalaminthos
cannabis	hemp	kannabis
capparis	caper	kapparis
capsella	shepherd's-purse	thlaspis
cardamine	bittercress	kardamis
cardamine	bittercress	kardamon
carthamus	safflower	atraktylis
carthamus	safflower	knekos
castanea	chestnut	kastanos
caucalis	burparsley	kaukalis
cedrus	cedar	kedros
centaurium	centaury	kentaurie
chaerophyllum	chervil	chairephyllon
chamaecyparis	groundcypress	chamaikyparissos
chamaemelum	chamomile	chamaimelon
chamaepeuce	groundlarch	chamaipeuke
chamaesyce	groundspurge	chamaisyke

109

chenopodium	goosefoot	osyris
cichorium	chicory, succory	kichoreion
cicuta	hemlock	koneion
circaea	enchanter's nightshade	kirkaia
cirsium	thistle	keanothos
cistus	rockrose	kisthos
clinopodium	basil	klinopodion
convolvulus	bindweed	doryknion
convolvulus	bindweed	iasione
convolvulus	bindweed	klymenon
convolvulus	bindweed	smilax
convolvulus	scammony	skamonia
conyza	fleabane	bakkaris
conyza	fleabane	konyza
cornus	dogwood	kranon
coronilla	crownvetch	hedysaron
cotinus	smokebush	thapsos
crataegus	hawthorn	krataigos
cressa	kidneyvetch	anthyllis
crithmum	samphire	krithmos
crocus	saffron	krokos
cucurbita	pumpkin	kolokynthe
cucurbita	cucumber	sikyos
cuminum	cummin	kyminon
cupressus	cypress	kyparissos
cuscuta	dodder	orobagche
cyclamen	sowbread	kyklamis
cydonia	quince	kydonia
cynara	globe artichoke	kinara
cynara	globe artichoke	skolymos
d. oleoides	dwarf-olive	chamelaea
delphinium	larkspur	delphinion
dipsacus	teazel	dipsakos
dipsacus	teazel	knaphos
draba	whitlowgrass	drabe
e. apios	acorn-spurge	chamaibalanos
ebenus	ebony	ebenos
echinops	globethistle	echinopous
echium	bugloss	echion
endivia	endive	seris
erica	heather	ereike
eu.peplis	purple spurge	peplis
eu.pithyusa	balearic spurge	pityoussa
euonymus	spindle	euonymos
euonymus	spindle-tree	tetragonia
euphorbia	spurge	tithymalos
f. sycomorus	fig-mulberry	sykomoros
fagus	beech	phegos
far	spelt	zeia

ferula	giant-fennel	narthex
ficus	fig	sykea
frangula	buckthorn	sphalax
frumentum	corn	olyra
frumentum	corn	stachys
galium	cleavers	aparine
genista	greenweed	chamaizelon
genista	broom	spartos
glechoma	groundivy	chamaikissos
gnaphalium	cudweed	gnaphalion
gnidium	healwood	thymelaea
gramen	grass	poa
grossus	winterfig	olynthos
hedera	ivy	kissos
hedera	ivy	skindapsos
helichrysum	everlasting	helichrysos
helleborus	hellebore	helleboros
helleborus	hellebore	skarphe
heracleum	cowparsnip	sphondylion
hieracium	hawkweed	hierakion
hieracium	hawkweed	sogchites
hordeum	barley	kachrys
hordeum	barley	krithe
horminum	clary	horminon
hypericum	St. John's wort	askyron
hyssopus	hyssop	hyssopos
ilex	holm-oak	prinos
insititia	bullace	spondias
iris	iris	iris
juncus	rush	thryon
juniperus	juniper	arkeuthos
juniperus	juniper	thua, thuia
kickxia	toadflax	elatine
lactuca	lettuce	thridakine
larix	larch	larix
laserpitium	laser	silphion
lathyrus	vetchling	lathyros
laurus	laurel	daphne
lavatera	treemallow	dendromalache
lentiscus	mastic	schinos
lepidium	pepperwort	iberis
ligustrum	privet	kelastra
lilium	lily	krinon
lilium	whitelily	leirion
linum	flax	linon
lolium	darnel	aira

lolium	darnel	thyaros
lolium	darnel	zizanion
lysimachia	loosestrife	lysimachion
malus	appletree	melea
malus	apple	melon
malva	muskmallow	alkea
malva	mallow	malache
mandragora	mandrake	mandragoras
marrubium	horehound	prasion
medicago	medick	kytisos
melampyrum	blackwheat	melampyron
mentha	mint	sisymbron
mespilus	medlar	mespile
meum	spignel	meon
milium	millet	kegchros
millefolium	milfoil	chiliophyllos
mimosa	wattle	aischunomene
mnium	moss	mnion
morus	black mulberry	moron
morus	white mulberry	sukaminos
muscus	moss	sphagnos
myriophyllum	watermilfoil	myriophyllon
myrrhis	sweet cicely	myrrhis
myrtus	myrtle	myrsine
myrtus	myrtle	myrtos
narcissus	narcissus	narkissos
nardus	nard	nardos
ocimum	basil	okimon
olea	olive	elaia
onobrychis	sainfoin	onobrychis
onopordon	cottonthistle	akanthion
onopordon	cottonthistle	onopordon
origanum	marjoram	knekion
origanum	marjoram	origanon
ornithogalum	Star of Bethlehem	ornithogalon
orobanche	broomrape	orobagche
orobus	bittervetch	orobos
ostrya	ironwood	ostrya
p. cerasus	dwarf cherry	chamaikerasos
paeonia	peony	paionia
paliurus	Christsthorn	paliouros
papaver	poppy	argemone
papaver	poppy	mekon
papaver	poppy	rhoias
papyrus	papersedge	byblos
parietaria	pellitory	helxine
parietaria	pellitory	pyrethron
peplis	purslane	andrachne
persica	peach	persike

petasites	butterbur	petasites
petroselinum	rockparsley	petroselinon
petroselinum	parsley	selinon
peucedanum	hogsfennel	peukedanon
phaseolus	bean	kyamos
phaseolus	kidneybean	phaselos
physalis	wintercherry	halikakabon
picea	spruce	peuke
pinus	pine	pitys
piper	pepper	peperi
pistacia	pistachio	pistake
pisum	pea	pisos
platanus	plane	platanos
populus	poplar	aigeiros
populus	poplar	kerkis
poterium	burnet	pheos
pr.amygdalus	almond	amygdalea
pr.cerasus	cherry	kerasos
prunus	sloe	brabylon
prunus	damson	brabylon
prunus	plum	proumne
punica	pomegranate	rhoa
pyrus	pear	achras
pyrus	pear	apion, apios
pyrus	pear	ogchne, onchne
qu.coccifera	kermes-oak	kokkos
qu.ilex	holm-oak	aria
qu.suber	cork-oak	phellos
quercus	oak	drys
ranunculus	buttercup	batrachion
rapa, rapum	turnip	rhapys, rhaphys
raphanus	radish	rhaphanis
reduvia	whitlow	paronychia
rhamnus	buckthorn	rhamnos
rhus	sumach	rhous
ricinus	castor	kiki
ricinus	castor	kroton
rosa	briar	rhachos
rosa	rose	rhodon
rosmarinus	rosemary	dendrolibanos
rosmarinus	rosemary	libanotis
rubia	madder	ereuthedanon
rubus	bramble	batos
rumex	sorrel	lapathon
ruscus	dwarf laurel	chamaidaphne
ruscus	butcher's broom	chamaimyrsine
ruscus	butchersbroom	hypoglosson
ruscus	butcher's broom	kentromyrsine
ruscus	butcher's broom	oxymyrsine
ruta	rue	peganon

113

s. ebulus	dwarf elder	chamaiakte
s. torminalis	service-tree	oa, oia, oua
salix	willow	itea
salvia	sage	sphakos
sambucus	elder	aktea
santalum	sandalwood	santalon
satureia	savory	thymbra
saxifraga	saxifrage	empetron
schoenus	rush	schoinos
scilla	squill	skilla
secale	rye	briza
sedum	stonecrop	anacampseros
senecio	groundsel	erigeron
sesamum	sesame	sesame
sesamum	sesame	sesamon
siligo	winterwheat	silignion
sinapis	mustard	sinapi
sium	waterparsnip	sion
smyrnium	alexanders	smyrnion
solanum	nightshade	strychnos
sonchus	sowthistle	sogchos
spartium	med.broom	spartion
statice	sealavender	statike
tamarix	tamarisk	myrike
taxus	yew	milos
terebinthus	turpentine-tree	terminthos
teucrium	germander	chamaidrys
teucrium	germander	skordion
teucrium	germander	teukrion
thalictrum	meadowrue	thaliktron
thelypteris	ladyfern	thelypteris
thymus	thyme	herpyllon
thymus	thyme	thymon
tilia	linden	philyra
tordylium	hartwort	seselis
tordylium	hartwort	tordylon
trigonella	fenugreek	telis
triticum	wheat	pyros
triticum	wheat	sitos
tus	frankincense	libanos
tussilago	coltsfoot	chamaigeiron
tussilago	coltsfoot	chamaileuke
tussilago	coltsfoot	petasites
ulmus	elm	ptelea
urtica	nettle	knide
verbascum	mullein	phlomos
viola	violet	ion
viscum	mistletoe	ixos
vitis	vine	ampelos

withania	sleepy nightshade	strychnon
xanthium	burrweed	xanthion
zingiber	ginger	ziggiberis
zizyphus	jujube	zizyphon

Three-language list of botanical name components

ENGLISH	LATIN	GREEK
acorn-spurge	e. apios	chamaibalanos
alexanders	smyrnium	smyrnion
almond	pr.amygdalus	amygdalea
apple	malus	melon
appletree	malus	melea
asarabacca	asarum	asaron
asparagus	asparagus	asparagos
balearic spurge	eu.pithyusa	pityoussa
barley	hordeum	kachrys
barley	hordeum	krithe
basil	clinopodium	klinopodion
basil	ocimum	okimon
bean	phaseolus	kyamos
beech	fagus	phegos
betony	betonica	prionitis
bindweed	convolvulus	doryknion
bindweed	convolvulus	iasione
bindweed	convolvulus	klymenon
bindweed	convolvulus	smilax
bittercress	cardamine	kardamis
bittercress	cardamine	kardamon
bittervetch	orobus	orobos
black mulberry	morus	moron
blackwheat	melampyrum	melampyron
box	buxus	pyxos
bramble	rubus	batos
briar	rosa	rhachos
broom	genista	spartos
broomrape	orobanche	orobagche
bryony	bryonia	kedrostis
buckthorn	rhamnus	rhamnos
buckthorn	frangula	sphalax
bugloss	echium	echion
bullace	insititia	spondias
burnet	poterium	pheos
burparsley	caucalis	kaukalis
burrweed	xanthium	xanthion
butcher's broom	ruscus	chamaimyrsine
butcher's broom	ruscus	kentromyrsine
butcher's broom	ruscus	oxymyrsine
butchersbroom	ruscus	hypoglosson
butterbur	petasites	petasites
buttercup	ranunculus	batrachion
cabbage	brassica	krambe
cane	bambusa	kanna
caper	capparis	kapparis
castor	ricinus	kiki
castor	ricinus	kroton

catmint	calamintha	kalaminthos
cedar	cedrus	kedros
centaury	centaurium	kentaurie
chamomile	chamaemelum	chamaimelon
cherry	pr.cerasus	kerasos
chervil	chaerophyllum	chairephyllon
chervil	anthriscus	skandix
chestnut	castanea	kastanos
chicory, succory	cichorium	kichoreion
Christsthorn	paliurus	paliouros
clary	horminum	horminon
cleavers	galium	aparine
coltsfoot	tussilago	petasites
coltsfoot	tussilago	chamaigeiron
coltsfopt	tussilago	chamaileuke
cork-oak	qu.suber	phellos
corn	frumentum	olyra
corn	frumentum	stachys
cottonthistle	onopordon	akanthion
cottonthistle	onopordon	onopordon
cowparsnip	heracleum	sphondylion
crownvetch	coronilla	hedysaron
cuckoopint	arum	arisaron
cucumber	cucurbita	sikyos
cudweed	gnaphalium	gnaphalion
cummin	cuminum	kyminon
cypress	cupressus	kyparissos
damson	prunus	brabylon
darnel	lolium	aira
darnel	lolium	thyaros
darnel	lolium	zizanion
dodder	cuscuta	orobagche
dogwood	cornus	kranon
dwarf cherry	p. cerasus	chamaikerasos
dwarf elder	s. ebulus	chamaiakte
dwarf laurel	ruscus	chamaidaphne
dwarf-olive	d. oleoides	chamelaea
ebony	ebenus	ebenos
elder	sambucus	aktea
elm	ulmus	ptelea
enchanter's nightshade	circaea	kirkaia
endive	endivia	seris
everlasting	helichrysum	helichrysos
fanpalm	borassus	borassos
fenugreek	trigonella	telis
fig	ficus	sykea
fig-mulberry	f. sycomorus	sykomoros
fir, spruce, pine	abies, picea, pinus	elate
flax	linum	linon

117

fleabane	conyza	bakkaris
fleabane	conyza	konyza
frankincense	tus	libanos
garlic	allium	skorodon
germander	teucrium	chamaidrys
germander	teucrium	skordion
germander	teucrium	teukrion
giant-fennel	ferula	narthex
ginger	zingiber	ziggiberis
globe artichoke	cynara	kinara
globe artichoke	cynara	skolymos
globethistle	echinops	echinopous
goosefoot	chenopodium	osyris
grass	gramen	poa
greenweed	genista	chamaizelon
groundcypress	chamaecyparis	chamaikyparissos
groundivy	glechoma	chamaikissos
groundlarch	chamaepeuce	chamaipeuke
groundpine	ajuga	chamaipitys
groundsel	senecio	erigeron
groundspurge	chamaesyce	chamaisyke
hartwort	tordylium	seselis
hartwort	tordylium	tordylon
hawkweed	hieracium	hierakion
hawkweed	hieracium	sogchites
hawthorn	crataegus	krataigos
healwood	gnidium	thymelaea
heather	erica	ereike
hellebore	helleborus	helleboros
hellebore	helleborus	skarphe
hemlock	cicuta	koneion
hemp	cannabis	kannabis
hogsfennel	peucedanum	peukedanon
holm-oak	qu.ilex	aria
holm-oak	ilex	prinos
horehound	ballota	ballote
horehound	marrubium	prasion
hyssop	hyssopus	hyssopos
iris	iris	iris
ironwood	ostrya	ostrya
ivy	hedera	kissos
ivy	hedera	skindapsos
jujube	zizyphus	zizyphon
juniper	juniperus	arkeuthos
juniper	juniperus	thua, thuia
kermes-oak	qu.coccifera	kokkos
kidneybean	phaseolus	phaselos

kidneyvetch	cressa	anthyllis
kingspear	asphodelus	asphodelos
ladyfern	thelypteris	thelypteris
larch	larix	larix
larkspur	delphinium	delphinion
laser	laserpitium	silphion
laurel	laurus	daphne
lettuce	lactuca	thridakine
lily	lilium	krinon
linden	tilia	philyra
loosestrife	lysimachia	lysimachion
love-lies-bleeding	amaranthus	amarantos
love-lies-bleeding	amaranthus	bliton
madder	rubia	ereuthedanon
mallow	malva	malache
mandrake	mandragora	mandragoras
maple	acer	sphendamnos
marjoram	origanum	knekion
marjoram	origanum	origanon
marshmallow	althaea	althaia
mastic	lentiscus	schinos
meadowrue	thalictrum	thaliktron
med.broom	spartium	spartion
medick	medicago	kytisos
medlar	mespilus	mespile
milfoil	millefolium	chiliophyllos
millet	milium	kegchros
mint	mentha	sisymbron
mistletoe	viscum	ixos
moss	mnium	mnion
moss	muscus	sphagnos
mullein	verbascum	phlomos
muskmallow	malva	alkea
mustard	sinapis	sinapi
myrtle	myrtus	myrsine
myrtle	myrtus	myrtos
narcissus	narcissus	narkissos
nard	nardus	nardos
nettle	urtica	knide
nightshade	solanum	strychnos
oak	quercus	drys
oat	avena	aigilops
oats	avena	bromos
olive	olea	elaia
onion	allium	kromyon
orache	atriplex	atraphaxis
papersedge	papyrus	byblos

119

parsley	petroselinum	selinon
pea	pisum	pisos
peach	persica	persike
pear	pyrus	achras
pear	pyrus	apion, apios
pear	pyrus	ogchne, onchne
pellitory	parietaria	helxine
pellitory	parietaria	pyrethron
peony	paeonia	paionia
pepper	piper	peperi
pepperwort	lepidium	iberis
pimpernel	anagallis	korchoros
pine	pinus	pitys
pistachio	pistacia	pistake
plane	platanus	platanos
plum	prunus	proumne
pomegranate	punica	rhoa
poplar	populus	aigeiros
poplar	populus	kerkis
poppy	papaver	argemone
poppy	papaver	mekon
poppy	papaver	rhoias
privet	ligustrum	kelastra
pumpkin	cucurbita	kolokynthe
purple spurge	eu.peplis	peplis
purslane	peplis	andrachne
quince	cydonia	kydonia
radish	raphanus	rhaphanis
reed	arundo	donax
rockparsley	petroselinum	petroselinon
rockrose	cistus	kisthos
rose	rosa	rhodon
rosemary	rosmarinus	dendrolibanos
rosemary	rosmarinus	libanotis
rue	ruta	peganon
rush	schoenus	schoinos
rush	juncus	thryon
rye	secale	briza
safflower	carthamus	atraktylis
safflower	carthamus	knekos
saffron	crocus	krokos
sage	salvia	sphakos
sainfoin	onobrychis	onobrychis
samphire	crithmum	krithmos
sandalwood	santalum	santalon
savory	satureia	thymbra
saxifrage	saxifraga	empetron
scammony	convolvulus	skamonia
sealavender	statice	statike

120

service-tree	s. torminalis	oa, oia, oua
sesame	sesamum	sesame
sesame	sesamum	sesamon
shepherd's-purse	capsella	thlaspis
sleepy nightshade	withania	strychnon
sloe	prunus	brabylon
smokebush	cotinus	thapsos
sorrel	rumex	lapathon
sowbread	cyclamen	kyklamis
sowthistle	sonchus	sogchos
spelt	far	zeia
spignel	meum	meon
spindle	euonymus	euonymos
spindle-tree	euonymus	tetragonia
spruce	picea	peuke
spurge	euphorbia	tithymalos
squill	scilla	skilla
St. John's wort	hypericum	askyron
Star of Bethlehem	ornithogalum	ornithogalon
stonecrop	sedum	anacampseros
strawberry tree	arbutus	komaros
sumach	rhus	rhous
sweet cicely	myrrhis	myrrhis
tamarisk	tamarix	myrike
teazel	dipsacus	dipsakos
teazel	dipsacus	knaphos
thistle	cirsium	keanothos
thyme	thymus	herpyllon
thyme	thymus	thymon
toadflax	kickxia	elatine
tragacanth	astragalus	aspalathos
tragacanth	astragalus	astragalos
treemallow	lavatera	dendromalache
turnip	rapa, rapum	rhapys, rhaphys
turpentine-tree	terebinthus	terminthos
tutsan	androsaemum	androsaimon
vetchling	lathyrus	lathyros
vine	vitis	ampelos
violet	viola	ion
watermilfoil	myriophyllum	myriophyllon
waterparsnip	sium	sion
wattle	mimosa	aischunomene
wheat	triticum	pyros
wheat	triticum	sitos
white mulberry	morus	sukaminos
whitelily	lilium	leirion
whitlow	reduvia	paronychia
whitlowgrass	draba	drabe
willow	salix	itea

Three-language list of botanical name components

wintercherry	physalis	halikakabon
winterfig	grossus	olynthos
winterwheat	siligo	silignion
wormwood	artemisia	seriphos
yarrow	achillea	ptarmike
yew	taxus	milos

NUMERICAL SUPPLEMENT

GREEK	LATIN	ENGLISH
amphi-	bi-	two
aniso-	inaequi-	odd
aperanto-	infini-	infinite
apeiro-	infini-	infinite
arithmo-	numeri-	number
chiliado-	mille-	thousand
chiliaki-	miliens	thousand x
chilio-	mille-	thousand
chiliokaipentekostaplasio-	millimodi et quinquagintuplici-	thousand & fiftyfold
chilioplasio-	millimodi-	thousandfold
chiliosto-	millensimi-	thousandth
deka-	decem	ten
dekache	deciens, decies	ten parts, into
dekachilio-	decem milia	ten thousand
dekaki-	deciens, decies	ten x
dekaplasio-	decemplici-	tenfold
dekaplasio-	decuplici-	tenfold
dekato-	decim-	tenth
deutero-	secundi-	second
di-	duplici-	twice
di-	dua-, duo-	two
diakopto-	bisecti-	two, cut in
diakosia-	ducenti-	two hundred
diakosiaki-	ducentiens, -ies	two hundred x
diakosiaplasio-	ducentuplici-	two hundred-fold
diatemno-	bisecti-	two, cut in
diatomo-	bisecti-	two, cut in
diatricho-	trivii-	three ways
diatrito-	tertiani-	tertian
dicha-, diche-, dicho-	bis	twice, in two
dichado-	dimidii-	half
dichado-	semi-	half
didymo-	gemini-	twin
dimero-, -e	bisecti-	two, cut in
diphasio-	bifarii-	two-rowed
diplaco-	duplici-	double
diplasiephemiso-	duo semis	two & a half
diplasiepidimero-	duo et duae partes	two & two-thirds
diplasiepidimoiro-	duo et duae partes	two & two-thirds
diplasiepiditrito-	duo et duae partes	two & two-thirds
diplasiepiekto-	duo et sexta pars	two & one-sixth
diplasiepipempto-	duo et quinta pars	two & one-fifth
diplasiepitetarto-	duo et quarta pars	two & one-quarter
diplasiepitetramero-, -e-	duo et quattuor partes	two & four-fifths
diplasiepitetrapempto-	duo et quattuor partes	two & four-fifths

diplasiepitrimero-, -e-	duo et tres partes	two & three-quarters
diplasiepitrito-	duo et triens	two & one-third
diplasio-	duplici-	double, twofold
diplo-	duplici-	double
dischilio-	duo milia	two thousand
dismyrio-	viginti milia	twenty thousand
disso-	duplici-	twofold
disticho-	bifarii-	two-rowed
dodeka-	duodecim	twelve
dodekaki-	duodeciens, -ies	twelve x
dodekakischilio-	duodecim milia	twelve thousand
dodekaplasio-	duodecuplici-	twelvefold
dodekasemo-	duodecimplici-	twelvefold
dodekato-	duodecim-	twelfth
dyasmo-	bis	twice, in two
dyo-	duo	two
dyodeka-	duodecim	twelve
dyokaideka-	duodecim	twelve
dyosto-	secundi-	second
enaki-, ennaki-	noviens, -ies	nine x
enakischilioi-	novem milia	nine thousand
enakosi-	nongenti	nine hundred
enakosiosto-	nongentensimi-	nine hundredth
enato-	noni-	ninth
enenekontaki-	nonagiens, -ies	ninety x
enenekonto-	nonaginta	ninety
enenekosto-	nonagensimi-	ninetieth
enna-, enne-	novem	nine
ennea-, enneado-	novem	nine
enneachilio-	novem milia	nine thousand
enneakaideka-	undeviginti	nineteen
enneakaidekaki-	undeviciens, -ies	nineteen x
enneakaidekaplasio-	undevigintiplici-	nineteenfold
enneakaidekato-	undevicensimi-	nineteenth
enneaki-	noviens, -ies	nine x
enneakischilio-	novem milia	nine thousand
enneakismyrio-	nonaginta milia	ninety thousand
enneaplasio-	novemplici-	ninefold
epiekto-	unus et sexta pars	one & one-sixth
epiogdoo-	unus et octava pars	one & one-eighth
epiogdoo-	sesquioctavi-	one & one-eighth
epipempto-	unus et quinta pars	one & one-fifth
epitetarto-	unus et quarta pars	one & one-quarter
epitetraebdomo-	unus et quater septima pars	one & four-sevenths
epitetramere-	unus et quattuor partes	one & four-fifths
epitetrapempto-	unus et quattuor partes	one & four-fifths
epitrimeri-	unus et tres partes	one & three-quarters
epitrito-	unus et tertia pars, sesquitertii-	one & one-third
haplo-	simplici-	single

hebdomado-	septem	seven
hebdomaki-	septiens, septies	seven x
hebdomato-	septimi-	seventh
hebdomekontaki-	septuagiens, -ies	seventy x
hebdomekonto-	septuaginta	seventy
hebdomekosto-	septuagensimi-	seventieth
hebdomo-	septimi-	seventh
hek-	sex	six
hekasto-	quisque	each one
hekatero-	alteruter	each of two
hekato-	centi-	hundred
hekatonta-	centi-	hundred
hekatontaki-	centiens, -ies	hundred x
hekatontaplasio-	centuplici-	hundredfold
hekatosto-	centensimi-	hundredth
hekkaideka-	sedecim	sixteen
hekkaidekaki-	sedeciens, -ies	sixteen x
hekkaidekato-	sextus decimi-	sixteenth
hekto-	sexti-	sixth
hemi-, hemisy-	semi-	half
hemioliasmo-	sesqui-	one & a half
hemiolo-	sesquialteri-	one & a half as much again
hemisytrito-	sesqui-	one & a half
hemisytritoplasio-	sesquiplici-	one & a half-fold
hendeka-	undecim	eleven
hendekaki-	undeciens, -ies	eleven x
hendekato-	undecimi-	eleventh
heniaio-	singuli-	single
heno-	uni-	one
hepta-, heptado-	septem	seven
heptacha-	septemfidi-	seven parts, into
heptakaideka	septendecim	seventeen
heptakaidekaki-	septiens deciens, -ies	seventeen x
heptakaidekato-	septimus decimi-	seventeenth
heptakaieikosamorio-	septem et vicensima pars	one-twenty-seventh
heptakaieikosaplasio-	septem et vigintiplici-	twenty-sevenfold
heptaki-	septiens	seven x
heptakischilio-	septem milia	seven thousand
heptakismyrio-	septuaginta milia	seventy thousand
heptakosio-	septingenti	seven hundred
heptakosioplasiaki-	septingentuplici-	seven hundred x
heptakosiosto-	septingentensimi-	seven hundredth
heptamoirio-, -morio-	septima pars	one-seventh
heptaplasio-, heptaploo-, -plou-	septuplici-	sevenfold
heptasemo-	septimetri-	seven times, of
hex-, hexa-	sex-	six
hexa-	sexti-	sixth
hexachei-, hexacha-	sexfidi-	six parts, into
hexaki-	sexiens, -ies	six x
hexakischilio-	sex milia	six thousand

hexakismyrio-	sexaginta milia	sixty thousand
hexakosi-	sescenti	six hundred
hexakosiosto-	sescentensimi-	six hundredth
hexamoro-, hexamoiro-	sexta pars, sextanti-	one-sixth
hexaplasio-, hexaple-,		
hexaploo-	sextuplici-	sixfold
hexe-	proximi-	next one
hexekonta-, hexekontado-	sexaginta	sixty
hexekontaki-	sexagiens, -ies	sixty x
hexekontamoiro-	sexagensima pars	one-sixtieth, of sixty parts
hexekosto-	sexagensimi-	sixtieth
hosaki-	totiens, -ies	as many x
hosaplasio-	tot multiplici-	as many fold
hypepitetarto-	tres partes, dodrans	three-quarters
hypepitrito-		
(fide L&S, non LSJ)	duae tertiae partes, bes	two-thirds
hypotetramere-	unus et quattuor partes	one & four-fifths
hypotetraplasiepitrito-	quattuor et tertia pars	four & one-third, less by
hypotetraplasio-	quarta pars, quadranti-	one-quarter
hypotriplasiepidipempto-	tres et bis quinta pars	three & two-fifths, less by
hypotrito-	tertia pars, trienti-	one-third, less by
hystero-	posteriori-	latter
ikosaki-	viciens, -ies	twenty x
ikosaplasio-	vigintiplici-	twentyfold
ikosi-, ikosa-	viginti	twenty
ikosiduo-	duo et viginti	twenty-two
iso-	aequi-	equal
meso-	semi-	half
mia- (f.)	uni-	one
mixo-	semi-	half
monacho-	solitarii-	solitary
monadiko-	solitarii-	solitary
monere-	singuli-	single
monia-	solitarii-	solitary
mono-	uni-	one
monoide-	uniformi-	uniform
monoteto-	uniti-	unity
myriado-	decem milia	ten thousand
myriaki-	deciens miliens, -ies	ten thousand x
myriakismyriosto-	centiens miliens milia	hundred millionth
myriadiko-	deciens millensimi-	ten thousandth
myrio-	innumerabili-	infinite
myrioplasio-	deciens millimodi-	ten thousandfold
ogdoadiko-	de octo	eight, of
ogdoado-	octo	eight
ogdoato-	octavi-	eighth
ogdoekonto-	octoginta	eighty
ogdoekosto-	octogensimi-	eightieth
ogdoo-	octavi-	eighth

oktacho-	octofidi-	eight ways, in
oktado-	octo-	eight
oktaki-	octiens, -ies	eight x
oktakischilio-	octo milia	eight thousand
oktakismyrio-	octoginta milia	eighty thousand
oktakosio-	octingenti	eight hundred
oktakosiosto-	octingentensimi-	eight hundredth
oktamere-	octopartiti-	eight parts, in
oktaplasio-	octupli-	eightfold
oktaploo-	octupli-	eightfold
okto-	octo-	eight
oktokaideka-	duodeviginti	eighteen
oktokaidekaki-	duodeviciens, -ies	eighteen x
oktokaidekaplasio-	duodevigintiplici-	eighteenfold
oktokaidekato-	duodevicensimi-	eighteenth
oktokaidekato-	octavus decimi-	eighteenth
oktokaieikosaplasio-	duodetrigintaplici-	twenty-eightfold
oligo-	pauci-	few
oligosto-	perpauci-	very few
oudenaki-	nulli-	zero x
pan-	toti-	all
pempe-	quinque	five
pempto-	quinti-	fifth
penta-, pente-	quinque	five
pentachilio-	quinque milia	five thousand
pentacho-	quinquefidi-	five ways, in
pentadiko-	de quinque	five, of
pentaki-	quinquiens, -ies	five x
pentakischilio-	quinque milia	five thousand
pentakischiliosto-	quinquiens millensimi-	five thousandth
pentakismyrio-	quinquaginta milia	fifty thousand
pentakosi-, pentekosi-	quingenti	five hundred
pentakosiosto-	quingentensimi-	five hundredth
pentamere-	quinquepartiti-	five-partite
pentaplasiephemiso-	quinque et dimidii-	five & a half
pentaplasiepipempto-	quinque et quinta pars	five & one-fifth
pentaplasiepitetarto-	quinque et quarta pars	five & a quarter
pentaplasiepitrito-	quinque et tertia pars	five & a third
pentaplasio-, -plesio-	quintuplici-	fivefold
pentaploo-	quintuplici-	fivefold
pentekaideka-	quindecim	fifteen
pentekaidekaki-	quindeciens, -ies	fifteen x
pentekaidekaplasio-	quindecimplici-	fifteenfold
pentekaidekato-	quintus decimi-	fifteenth
pentekaieikosi-	quinque et viginti	twenty-five
pentekaieikosto-	quinque et vicensimi-	twenty-fifth
pentekonta-	quinquaginta	fifty
pentekontaki-	quinquagiens, -ies	fifty x
pentekontaplasio-	quinquagintuplici-	fiftyfold
pentekosto-	quinquagensimi-	fiftieth
penthemimere-	duo et dimidii-	two & half

pentongkio-, pentoungkio-	quincunci-	quincunx
pleio-	pluri-	more
pleisto-	plurimi-	most
pollaki-, pollache-	saepe	many x, often
pollakismyrio-	infiniti-	many ten thousands
pollaplasio-	multiplici-	many x more
pollaploo-	multiplici-	manifold
pollosto-	multesimi-	one of many
poly-	multi-	many, much
posaki-	quotiens, -ies	how often?
protisto-	perprimi-	very first, the
proto-	primi-	first
spithame-	dodranti-	span
tessara-	quattuor	four
tessarakaideka-	quattuordecim	fourteen
tessarakaidekaki-	quattuordeciens, -ies	fourteen x
tessarakaidekaplasio-	quattuordecimplici-	fourteenfold
tessarakaidekato-	quartus decimi-	fourteenth
tessarakonta-	quadraginta	forty
tessarakontaki-	quadragiens, -ies	forty x
tessarakontaplasio-	quadragintuplici-	fortyfold
tessarakosto-	quadragensimi-	fortieth
tessarakostogdoo-	duodequinquagensimi-	forty-eighth
tessareskaideka-	quattuordecim	fourteen
tetarto-, tetranto-, tetrato-	quarti-	fourth
tetra-, tetrado-	quattuor	four
tetracha-	quadrifidi-	four ways, in
tetradymo-, tetrazygo-	quadruplici-	fourfold
tetraki-	quater	four x
tetrakischilio-	quattuor milia	four thousand
tetrakismyrio-	quadraginta milia	forty thousand
tetrakosio-	quadringenti	four hundred
tetrakosiosto-	quadringentensimi-	four hundredth
tetramere-	quadripartiti-	four-partite
tetramoiro-	quadruplici-	fourfold
tetraplasiephemiso-	quattuor et dimidii-	four & a half
tetraplasiepidimere-	quattuor et duae partes	four & two-thirds
tetraplasiepipempto-	quattuor et quinta pars	four & one-fifth
tetraplasiepitetarto-	quattuor et quarta pars	four & a quarter
tetraplasiepitetramere-	quattuor et quattuor partes	four & four-fifths
tetraplasiepitrimere-	quattuor et tres partes	four & three-quarters
tetraplasiepitrito-	quattuor et tertia pars	four & a third
tetraplasio-, tetraploo-	quadruplici-	fourfold
tetraptycho-	quadruplici-	fourfold
tetraschisto-, tetratomo-	quadripartiti-	four-partite
tosaki-	totiens, -ies	so many x, so often
trei-	tres, tri-	three
treiskaideka-, triskaideka-	tredecim	thirteen
tri-, triado-	tres, tri-	three
triadiko-	triplici-	threefold, triple

triakado-, triakonto-	triginta	thirty
triaki-	ter	thrice, three x
triakontaki-	triciens, -ies	thirty x
triakontamorio-	tricensima pars	thirtieth part
triakontaplasio-	trigintuplici-	thirtyfold
triakosio-	trecenti	three hundred
triakosiochoo-	trecentuplici-	three hundredfold
triakostamorio-	tricensima pars	thirtieth part
triakosto-	tricensimi-	thirtieth
triakostodyo-	alter et tricensimi-	thirty-second
triakostopempto-	quinque et tricensimi-	thirty-fifth
tricha-, triche-, tricho-	trifarii-	three, in
trichtha-	tripartiti-	tripartite
trichthadio-	triplici-	threefold, triple
tridymo-, trizygo-	triplici-	threefold, triple
trikro-	trifidi-	trifid
trimere-, trimoiro-	tripartiti-	tripartite
trimoiriaio-	tres partes	three-quarters
triphasio-, triphato-	triplici-	threefold, triple
triplaco-, triploci-, triploo-	triplici-	threefold, triple
triplasiephebdomo-	tres et septima pars	three & one-seventh
triplasiephemiso-	tres et dimidii-	three & a half
triplasiepidimere-	tres et duae partes	three & two-thirds
triplasiepipempto-	tres et quinta pars	three & one-fifth
triplasiepitetarto-	tres et quarta pars	three & a quarter
triplasiepitetramere-	tres et quattuor partes	three & four-fifths
triplasiepitrimere-	tres et tres partes	three & three-quarters
triplasiepitrito-	tres et tertia pars	three & a third
triplasio-	triplici-	threefold, triple
tris-	ter	three x, thrice
trischide-, trischisto-	trifidi-	trifid, three-cleft
trischilio-	tria milia	three thousand
trischiliosto-	ter millensimi-	three thousandth
trischiliotrismyrio-	tres et triginta milia	thirty-three thousand
triskaidekaki-	terdeciens, -ies	thirteen x
triskaidekaplasio-	tredecimplici-	thirteenfold
triskaidekato-	tertius decimi-	thirteenth
trismyrio-	triginta milia	thirty thousand
trismyrioplasio-	triginta millimodi-	thirty thousandfold
trismyriosto-	triciens millensimi-	thirty thousandth
trissaki-	ter	thrice, three x
trisso-, tritto-, trixo-	triplici-	threefold, triple
tritemorio-	trieni-	one-third
trito-	tertii-	third
trittuo-, tritu-, trittua-	ternii-	three, the no.

LATIN	ENGLISH	GREEK
aequi-	equal	iso-
alter et tricensimi-	thirty-second	triakostodyo-
alteruter	each of two	hekatero-
bes, duae tertiae partes	two-thirds	hypepitrito- (fide L&S, non LSJ)
bi-	two	amphi-
bifarii-	two-rowed	diphasio-
bifarii-	two-rowed	disticho-
bis	twice, in two	dicha-, diche-, dicho-
bis	twice, in two	dyasmo-
bisecti-	two, cut in	diakopto-
bisecti-	two, cut in	diatemno-
bisecti-	two, cut in	diatomo-
bisecti-	two, cut in	dimero-, -e
centensimi-	hundredth	hekatosto-
centi-	hundred	hekato-
centi-	hundred	hekatonta-
centiens, -ies	hundred x	hekatontaki-
centiens miliens milia	hundred millionth	myriakismyriosto-
centuplici-	hundredfold	hekatontaplasio-
de octo	eight, of	ogdoadiko-
de quinque	five, of	pentadiko-
decem	ten	deka-
decem milia	ten thousand	dekachilio-
decem milia	ten thousand	myriado-
decemplici-	tenfold	dekaplasio-
deciens	ten parts, into	dekache
deciens	ten x	dekaki-
deciens miliens, -ies	ten thousand x	myriaki-
deciens millensimi-	ten thousandth	myriadiko-
deciens millimodi-	ten thousandfold	myrioplasio-
decim-	tenth	dekato-
decuplici-	tenfold	dekaplasio-
dimidii-	half	dichado-
dodranti-	span	spithame-
dua-, duo-	two	di-
duae partes, bes	two-thirds	hypepitrito- (fide L&S, non LSJ)
ducenti-	two hundred	diakosia-
ducentiens, -ies	two hundred x	diakosiaki-
ducentuplici-	two hundred-fold	diakosiaplasio-
duo	two	dyo-
duo et dimidii-	two & half	penthemimere-
duo et duae partes	two & two-thirds	diplasiepidimero-
duo et duae partes	two & two-thirds	diplasiepidimoiro-
duo et duae partes	two & two-thirds	diplasiepiditrito-
duo et quarta pars	two & one-quarter	diplasiepitetarto-

duo et quattuor partes	two & four-fifths	diplasiepitetramero-
duo et quattuor partes	two & four-fifths	diplasiepitetrapempto-
duo et quinta pars	two & one-fifth	diplasiepipempto-
duo et sexta pars	two & one-sixth	diplasiepiekto-
duo et tres partes	two & three-quarters	diplasiepitrimero-
duo et triens, -ies	two & one-third	diplasiepitrito-
duo et viginti	twenty-two	ikosiduo-
duo milia	two thousand	dischilio-
duo semis	two & a half	diplasiephemiso-
duodeciens, -ies	twelve x	dodekaki-
duodecim	twelve	dodeka-
duodecim	twelve	dyodeka-
duodecim	twelve	dyokaideka-
duodecim milia	twelve thousand	dodekakischilio-
duodecim-	twelfth	dodekato-
duodecimplici-	twelvefold	dodekasemo-
duodecuplici-	twelvefold	dodekaplasio-
duodequinquagensimi-	forty-eighth	tessarakostogdoo-
duodetrigintaplici-	twenty-eightfold	oktokaieikosaplasio-
duodevicensimi-	eighteenth	oktokaidekato-
duodeviciens, -ies	eighteen x	oktokaidekaki-
duodeviginti	eighteen	oktokaideka-
duodevigintiplici-	eighteenfold	oktokaidekaplasio-
duplici-	twice	di-
duplici-	double	diplaco-
duplici-	double, twofold	diplasio-
duplici-	double	diplo-
duplici-	twofold	disso-
gemini-	twin	didymo-
inaequi-	odd	aniso-
infini-	infinite	aperanto-
infini-	infinite	apeiro-
infiniti-	many ten thousands	pollakismyrio-
innumerabili-	infinite	myrio-
miliens, -ies	thousand x	chiliaki-
mille-	thousand	chiliado-
mille-	thousand	chilio-
millensimi-	thousandth	chiliosto-
millimodi et		
quinquagintuplici-	thousand & fiftyfold	chiliokaipentekostaplasio-
millimodi-	thousandfold	chilioplasio-
multesimi-	one of many	pollosto-
multi-	many, much	poly-
multiplici-	many x more	pollaplasio-
multiplici-	manifold	pollaploo-
nonagensimi-	ninetieth	enenekosto-
nonagiens, -ies	ninety x	enenekontaki-
nonaginta	ninety	enenekonto-

nonaginta milia	ninety thousand	enneakismyrio-
nongentensimi-	nine hundredth	enakosiosto-
nongenti	nine hundred	enakosi-
noni-	ninth	enato-
novem	nine	enna-, enne-
novem	nine	ennea-, enneado-
novem milia	nine thousand	enakischilioi-
novem milia	nine thousand	enneachilio-
novem milia	nine thousand	enneakischilio-
novemplici-	ninefold	enneaplasio-
noviens, -ies	nine x	enaki-, ennaki-
noviens, -ies	nine x	enneaki-
nulli-	zero x	oudenaki-
numeri-	number	arithmo-
octavi-	eighth	ogdoato-
octavi-	eighth	ogdoo-
octavus decimi-	eighteenth	oktokaidekato-
octiens, -ies	eight x	oktaki-
octingentensimi-	eight hundredth	oktakosiosto-
octingenti	eight hundred	oktakosio-
octo	eight	ogdoado-
octo milia	eight thousand	oktakischilio-
octo-	eight	oktado-
octo-	eight	okto-
octofidi-	eight ways, in	oktacho-
octogensimi-	eightieth	ogdoekosto-
octoginta	eighty	ogdoekonto-
octoginta milia	eighty thousand	oktakismyrio-
octopartiti-	eight parts, in	oktamere-
octupli-	eightfold	oktaplasio-
octupli-	eightfold	oktaploo-
pauci-	few	oligo-
perpauci-	very few	oligosto-
perprimi-	very first, the	protisto-
pluri-	more	pleio-
plurimi-	most	pleisto-
posteriori-	latter	hystero-
primi-	first	proto-
proximi-	next one	hexe-
quadragensimi-	fortieth	tessarakosto-
quadragiens, -ies	forty x	tessarakontaki-
quadraginta	forty	tessarakonta-
quadraginta milia	forty thousand	tetrakismyrio-
quadragintuplici-	fortyfold	tessarakontaplasio-
quadrifidi-	four ways, in	tetracha-
quadringentensimi-	four hundredth	tetrakosiosto-
quadringenti	four hundred	tetrakosio-
quadripartiti-	four-partite	tetramere-
quadripartiti-	four-partite	tetraschisto-, tetratomo-

quadruplici-	fourfold	tetradymo-, tetrazygo-
quadruplici-	fourfold	tetramoiro-
quadruplici-	fourfold	tetraplasio-, tetraploo-
quadruplici-	fourfold	tetraptycho-
quarta pars, quadranti-	one-quarter	hypotetraplasio-
quarti-	fourth	tetarto-, tetranto-, tetrato-
quartus decimi-	fourteenth	tessarakaidekato-
quater	four x	tetraki-
quattuor	four	tessara-
quattuor	four	tetra-, tetrado-
quattuor et quattuor partes	four & four-fifths	tetraplasiepitetramere-
quattuor et dimidii-	four & a half	tetraplasiephemiso-
quattuor et duae partes	four & two-thirds	tetraplasiepidimere-
quattuor et quarta pars	four & a quarter	tetraplasiepitetarto-
quattuor et quinta pars	four & one-fifth	tetraplasiepipempto-
quattuor et tertia pars	four & one-third, less by	hypotetraplasiepitrito-
quattuor et tertia pars	four & a third	tetraplasiepitrito-
quattuor et tres partes	four & three-quarters	tetraplasiepitrimere-
quattuor milia	four thousand	tetrakischilio-
quattuordeciens, -ies	fourteen x	tessarakaidekaki-
quattuordecim	fourteen	tessarakaideka-
quattuordecim	fourteen	tessareskaideka-
quattuordecimplici-	fourteenfold	tessarakaidekaplasio-
quincunci-	quincunx	pentongkio-, pentoungkio-
quindeciens, -ies	fifteen x	pentekaidekaki-
quindecim	fifteen	pentekaideka-
quindecimplici-	fifteenfold	pentekaidekaplasio-
quingentensimi-	five hundredth	pentakosiosto-
quingenti	five hundred	pentakosi-, pentekosi-
quinquagensimi-	fiftieth	pentekosto-
quinquagiens, -ies	fifty x	pentekontaki-
quinquaginta	fifty	pentekonta-
quinquaginta milia	fifty thousand	pentakismyrio-
quinquagintuplici-	fiftyfold	pentekontaplasio-
quinque	five	pempe-
quinque	five	penta-, pente-
quinque et dimidii-	five & a half	pentaplasiephemiso-
quinque et quarta pars	five & a quarter	pentaplasiepitetarto-
quinque et quinta pars	five & one-fifth	pentaplasiepipempto-
quinque et tertia pars	five & a third	pentaplasiepitrito-
quinque et tricensimi-	thirty-fifth	triakostopempto-
quinque et vicensimi-	twenty-fifth	pentekaieikosto-
quinque et viginti	twenty-five	pentekaieikosi-
quinque milia	five thousand	pentachilio-
quinque milia	five thousand	pentakischilio-
quinquefidi-	five ways, in	pentacho-
quinquepartiti-	five-partite	pentamere-
quinquiens, -ies	five x	pentaki-
quinquiens millensimi-	five thousandth	pentakischiliosto-
quinti-	fifth	pempto-
quintuplici-	fivefold	pentaplasio-, -plesio-
quintuplici-	fivefold	pentaploo-

quintus decimi-	fifteenth	pentekaidekato-
quisque	each one	hekasto-
quotiens, -ies	how often?	posaki-
saepe	many x, often	pollaki-, pollache-
secundi-	second	deutero-
secundi-	second	dyosto-
sedeciens, -ies	sixteen x	hekkaidekaki-
sedecim	sixteen	hekkaideka-
semi-	half	dichado-
semi-	half	hemi-, hemisy-
semi-	half	meso-
semi-	half	mixo-
septem	seven	hebdomado-
septem	seven	hepta-, heptado-
septem et vicensima pars	one-twenty-seventh	heptakaieikosamorio-
septem et vigintiplici-	twenty-sevenfold	heptakaieikosaplasio-
septem milia	seven thousand	heptakischilio-
septemfidi-	seven parts, into	heptacha-
septendecim	seventeen	heptakaideka
septiens, -ies	seven x	hebdomaki-
septiens, -ies	seven x	heptaki-
septiens deciens, -ies	seventeen x	heptakaidekaki-
septima pars	one-seventh	heptamoirio-, -morio-
septimetri-	seven times, of	heptasemo-
septimi-	seventh	hebdomato-
septimi-	seventh	hebdomo-
septimus decimi-	seventeenth	heptakaidekato-
septingentensimi-	seven hundredth	heptakosiosto-
septingenti	seven hundred	heptakosio-
septingentuplici-	seven hundred x	heptakosioplasiaki-
septuagensimi-	seventieth	hebdomekosto-
septuagiens, -ies	seventy x	hebdomekontaki-
septuaginta	seventy	hebdomekonto-
septuaginta milia	seventy thousand	heptakismyrio-
septuplici-	sevenfold	heptaplasio-, heptaploo-, -plou-
sescentensimi-	six hundredth	hexakosiosto-
sescenti	six hundred	hexakosi-
sesqui-	one & a half	hemioliasmo-
sesqui-	one & a half	hemisytrito-
sesquialteri-	one & a half as much again	hemiolo-
sesquioctavi-	one & one-eighth	epiogdoo-
sesquiplici-	one & a half-fold	hemisytritoplasio-
sex	six	hek-
sex milia	six thousand	hexakischilio-
sex-	six	hex-, hexa-
sexagensima pars	one-sixtieth, of sixty parts	hexekontamoiro-
sexagensimi-	sixtieth	hexekosto-
sexagiens, -ies	sixty x	hexekontaki-
sexaginta	sixty	hexekonta-, hexekontado-

134

sexaginta milia	sixty thousand	hexakismyrio-
sexfidi-	six parts, into	hexachei-, hexacha-
sexiens, -ies	six x	hexaki-
sexta pars, sextanti-	one-sixth	hexamoro-, hexamoiro-
sexti-	sixth	hekto-
sexti-	sixth	hexa-
sextuplici-	sixfold	hexaplasio-, hexaple-
sextuplici-	sixfold	hexaploo-
sextus decimi-	sixteenth	hekkaidekato-
simplici-	single	haplo-
singuli-	single	heniaio-
singuli-	single	monere-
solitarii-	solitary	monacho-
solitarii-	solitary	monadiko-
solitarii-	solitary	monia-
ter	thrice, three x	triaki-
ter	three x, thrice	tris-
ter	thrice, three x	trissaki-
ter millensimi-	three thousandth	trischiliosto-
terdeciens, -ies	thirteen x	triskaidekaki-
ternii-	three, the no.	trittuo-, tritu-, trittua-
tertia pars, trienti-	one-third, less by	hypotrito-
tertiani-	tertian	diatrito-
tertii-	third	trito-
tertius decimi-	thirteenth	triskaidekato-
tot multiplici-	as many fold	hosaplasio-
toti-	all	pan-
totiens, -ies	as many x	hosaki-
totiens, -ies	so many x, so often	tosaki-
trecenti	three hundred	triakosio-
trecentuplici-	three hundredfold	triakosiochoo-
tredecim	thirteen	treiskaideka-, triskaideka-
tredecimplici-	thirteenfold	triskaidekaplasio-
tres et bis quinta pars	three & two-fifths, less by	hypotriplasiepidipempto-
tres et dimidii-	three & a half	triplasiephemiso-
tres et duae partes	three & two-thirds	triplasiepidimere-
tres et quarta pars	three & a quarter	triplasiepitetarto-
tres et quattuor partes	three & four-fifths	triplasiepitetramere-
tres et quinta pars	three & one-fifth	triplasiepipempto-
tres et septima pars	three & one-seventh	triplasiephebdomo-
tres et tertia pars	three & a third	triplasiepitrito-
tres et tres partes	three & three-quarters	triplasiepitrimere-
tres et triginta milia	thirty-three thousand	trischiliotrismyrio-
tres partes	three-quarters	trimoiriaio-
tres partes, dodrans	three-quarters	hypepitetarto-
tres, tri-	three	trei-
tres, tri-	three	tri-, triado-
tria milia	three thousand	trischilio-
tricensima pars	thirtieth part	triakontamorio-
tricensima pars	thirtieth part	triakostamorio-
tricensimi-	thirtieth	triakosto-

triciens, -ies	thirty x	triakontaki-
triciens millensimi-	thirty thousandth	trismyriosto-
trieni-	one-third	tritemorio-
trifarii-	three, in	tricha-, triche-, tricho-
trifidi-	trifid	trikro-
trifidi-	trifid, three-cleft	trischide-, trischisto-
triginta	thirty	triakado-, triakonto-
triginta milia	thirty thousand	trismyrio-
triginta millimodi-	thirty thousandfold	trismyrioplasio-
trigintuplici-	thirtyfold	triakontaplasio-
tripartiti-	tripartite	trichtha-
tripartiti-	tripartite	trimere-, trimoiro-
triplici-	threefold, triple	triadiko-
triplici-	threefold, triple	trichthadio-
triplici-	threefold, triple	tridymo-, trizygo-
triplici-	threefold, triple	triphasio-, triphato-
triplici-	threefold, triple	triplaco-, triploci-, triploo-
triplici-	threefold, triple	triplasio-
triplici-	threefold, triple	trisso-, tritto-, trixo-
trivii-	three ways	diatricho-
undeciens, -ies	eleven x	hendekaki-
undecim	eleven	hendeka-
undecimi-	eleventh	hendekato-
undevicensimi-	nineteenth	enneakaidekato-
undeviciens, -ies	nineteen x	enneakaidekaki-
undeviginti	nineteen	enneakaideka-
undevigintiplici-	nineteenfold	enneakaidekaplasio-
uni-	one	heno-
uni-	one	mia- (f.)
uni-	one	mono-
uniformi-	uniform	monoide-
uniti-	unity	monoteto-
unus et octava pars	one & one-eighth	epiogdoo-
unus et quarta pars	one & one-quarter	epitetarto-
unus et quater septima pars	one & four-sevenths	epitetraebdomo-
unus et quattuor partes	one & four-fifths	epitetrapempto-
unus et quattuor partes	one & four-fifths	epitetramere-
unus et quattuor partes	one & four-fifths	hypotetramere-
unus et quinta pars	one & one-fifth	epipempto-
unus et sexta pars	one & one-sixth	epiekto-
unus et tertia pars, sesquitertii-	one & one-third	epitrito-
unus et tres partes	one & three-quarters	epitrimeri-
viciens, -ies	twenty x	ikosaki-
viginti	twenty	ikosi-, ikosa-
viginti milia	twenty thousand	dismyrio-
vigintiplici-	twentyfold	ikosaplasio-

ENGLISH	LATIN	GREEK
all	toti-	pan-
as many fold	tot multiplici-	hosaplasio-
as many x	totiens, -ies	hosaki-
double	duplici-	diplaco-
double	duplici-	diplo-
double, twofold	duplici-	diplasio-
each of two	alteruter	hekatero-
each one	quisque	hekasto-
eight	octo	ogdoado-
eight	octo-	oktado-
eight	octo-	okto-
eight hundred	octingenti	oktakosio-
eight hundredth	octingentensimi-	oktakosiosto-
eight parts, in	octopartiti-	oktamere-
eight thousand	octo milia	oktakischilio-
eight ways, in	octofidi-	oktacho-
eight x	octiens, -ies	oktaki-
eight, of	de octo	ogdoadiko-
eighteen	duodeviginti	oktokaideka-
eighteen x	duodeviciens, -ies	oktokaidekaki-
eighteenfold	duodevigintiplici-	oktokaidekaplasio-
eighteenth	duodevicensimi-	oktokaidekato-
eighteenth	octavus decimi-	oktokaidekato-
eightfold	octupli-	oktaplasio-
eightfold	octupli-	oktaploo-
eighth	octavi-	ogdoato-
eighth	octavi-	ogdoo-
eightieth	octogensimi-	ogdoekosto-
eighty	octoginta	ogdoekonto-
eighty thousand	octoginta milia	oktakismyrio-
eleven	undecim	hendeka-
eleven x	undeciens, -ies	hendekaki-
eleventh	undecimi-	hendekato-
equal	aequi-	iso-
few	pauci-	oligo-
fifteen	quindecim	pentekaideka-
fifteen x	quindeciens, -ies	pentekaidekaki-
fifteenfold	quindecimplici-	pentekaidekaplasio-
fifteenth	quintus decimi-	pentekaidekato-
fifth	quinti-	pempto-
fiftieth	quinquagensimi-	pentekosto-
fifty	quinquaginta	pentekonta-
fifty thousand	quinquaginta milia	pentakismyrio-
fifty x	quinquagiens, -ies	pentekontaki-
fiftyfold	quinquagintuplici-	pentekontaplasio-
first	primi-	proto-
five	quinque	pempe-

five	quinque	penta-, pente-
five & a half	quinque et dimidii-	pentaplasiephemiso-
five & a quarter	quinque et quarta pars	pentaplasiepitetarto-
five & a third	quinque et tertia pars	pentaplasiepitrito-
five & one-fifth	quinque et quinta pars	pentaplasiepipempto-
five hundred	quingenti	pentakosi-, pentekosi-
five hundredth	quingentensimi-	pentakosiosto-
five thousand	quinque milia	pentachilio-
five thousand	quinque milia	pentakischilio-
five thousandth	quinquiens millensimi-	pentakischiliosto-
five ways, in	quinquefidi-	pentacho-
five x	quinquiens, -ies	pentaki-
five, of	de quinque	pentadiko-
five-partite	quinquepartiti-	pentamere-
fivefold	quintuplici-	pentaplasio-, -plesio-,
		pentaploo-
fortieth	quadragensimi-	tessarakosto-
forty	quadraginta	tessarakonta-
forty thousand	quadraginta milia	tetrakismyrio-
forty x	quadragiens, -ies	tessarakontaki-
forty-eighth	duodequinquagensimi-	tessarakostogdoo-
fortyfold	quadragintuplici-	tessarakontaplasio-
four	quattuor	tessara-
four	quattuor	tetra-, tetrado-
four & a half	quattuor et dimidii-	tetraplasiephemiso-
four & a quarter	quattuor et quarta pars	tetraplasiepitetarto-
four & a third	quattuor et tertia pars	tetraplasiepitrito-
four & four-fifths	quattuor et quattuor partes	tetraplasiepitetramere-
four & one-fifth	quattuor et quinta pars	tetraplasiepipempto-
four & one-third, less by	quattuor et tertia pars	hypotetraplasiepitrito-
four & three-quarters	quattuor et tres partes	tetraplasiepitrimere-
four & two-thirds	quattuor et duae partes	tetraplasiepidimere-
four hundred	quadringenti	tetrakosio-
four hundredth	quadringentensimi-	tetrakosiosto-
four thousand	quattuor milia	tetrakischilio-
four ways, in	quadrifidi-	tetracha-
four x	quater	tetraki-
four-partite	quadripartiti-	tetramere-
four-partite	quadripartiti-	tetraschisto-, tetratomo-
fourfold	quadruplici-	tetradymo-, tetrazygo-
fourfold	quadruplici-	tetramoiro-
fourfold	quadruplici-	tetraplasio-, tetraploo-
fourfold	quadruplici-	tetraptycho-
fourteen	quattuordecim	tessarakaideka-
fourteen	quattuordecim	tessareskaideka-
fourteen x	quattuordeciens, -ies	tessarakaidekaki-
fourteenfold	quattuordecimplici-	tessarakaidekaplasio-
fourteenth	quartus decimi-	tessarakaidekato-
fourth	quarti-	tetarto-, tetranto-, tetrato-
half	dimidii-	dichado-
half	semi-	dichado-

half	semi-	hemi-, hemisy-
half	semi-	meso-
half	semi-	mixo-
how often?	quotiens, -ies	posaki-
hundred	centi-	hekato-
hundred	centi-	hekatonta-
hundred millionth	centiens miliens milia	myriakismyriosto-
hundred x	centiens, -ies	hekatontaki-
hundredfold	centuplici-	hekatontaplasio-
hundredth	centensimi-	hekatosto-
infinite	infini-	aperanto-
infinite	infini-	apeiro-
infinite	innumerabili-	myrio-
latter	posteriori-	hystero-
manifold	multiplici-	pollaploo-
many ten thousands	infiniti-	pollakismyrio-
many x more	multiplici-	pollaplasio-
many x, often	saepe	pollaki-, pollache-
many, much	multi-	poly-
more	pluri-	pleio-
most	plurimi-	pleisto-
next one	proximi-	hexe-
nine	novem	enna-, enne-, ennea-, enneado-
nine hundred	nongenti	enakosi-
nine hundredth	nongentensimi-	enakosiosto-
nine thousand	novem milia	enakischilioi-
nine thousand	novem milia	enneachilio-
nine thousand	novem milia	enneakischilio-
nine x	noviens, -ies	enaki-, ennaki-
nine x	noviens, -ies	enneaki-
ninefold	novemplici-	enneaplasio-
nineteen	undeviginti	enneakaideka-
nineteen x	undeviciens, -ies	enneakaidekaki-
nineteenfold	undevigintiplici-	enneakaidekaplasio-
nineteenth	undevicensimi-	enneakaidekato-
ninetieth	nonagensimi-	enenekosto-
ninety	nonaginta	enenekonto-
ninety thousand	nonaginta milia	enneakismyrio-
ninety x	nonagiens, -ies	enenekontaki-
ninth	noni-	cnato-
number	numeri-	arithmo-
odd	inaequi-	aniso-
one	uni-	heno-
one	uni-	mia- (f.)
one	uni-	mono-
one & a half	sesqui-	hemioliasmo-

one & a half	sesqui-	hemisytrito-
one & a half as much again	sesquialteri-	hemiolo-
one & a half-fold	sesquiplici-	hemisytritoplasio-
one & four-fifths	unus et quattuor partes	epitetrapempto-, epitetramere-
one & four-fifths	unus et quattuor partes	hypotetramere-
one & four-sevenths	unus et quater septima pars	epitetraebdomo-
one & one-eighth	sesquioctavi-	epiogdoo-
one & one-eighth	unus et octava pars	epiogdoo-
one & one-fifth	unus et quinta pars	epipempto-
one & one-quarter	unus et quarta pars	epitetarto-
one & one-sixth	unus et sexta pars	epiekto-
one & one-third	unus et tertia pars, sesquitertii-	epitrito-
one & three-quarters	unus et tres partes	epitrimeri-
one of many	multesimi-	pollosto-
one-quarter	quarta pars, quadranti-	hypotetraplasio-
one-seventh	septima pars	heptamoirio-, -morio-
one-sixth	sexta pars, sextanti-	hexamoro-, hexamoiro-
one-sixtieth, of sixty parts	sexagensima pars	hexekontamoiro-
one-third	trieni-	tritemorio-
one-third, less by	tertia pars, trienti-	hypotrito-
one-twenty-seventh	septem et vicensima pars	heptakaieikosamorio-
quincunx	quincunci-	pentongkio-, pentoungkio-
second	secundi-	deutero-
second	secundi-	dyosto-
seven	septem	hebdomado-
seven	septem	hepta-, heptado-
seven hundred	septingenti	heptakosio-
seven hundred x	septingentuplici-	heptakosioplasiaki-
seven hundredth	septingentensimi-	heptakosiosto-
seven parts, into	septemfidi-	heptacha-
seven thousand	septem milia	heptakischilio-
seven times, of	septimetri-	heptasemo-
seven x	septiens, -ies	hebdomaki-
seven x	septiens, -ies	heptaki-
sevenfold	septuplici-	heptaplasio-, heptaploo-, -plou-
seventeen	septendecim	heptakaideka
seventeen x	septiens deciens, -ies	heptakaidekaki-
seventeenth	septimus decimi-	heptakaidekato-
seventh	septimi-	hebdomato-
seventh	septimi-	hebdomo-
seventieth	septuagensimi-	hebdomekosto-
seventy	septuaginta	hebdomekonto-
seventy thousand	septuaginta milia	heptakismyrio-
seventy x	septuagiens, -ies	hebdomekontaki-
single	simplici-	haplo-
single	singuli-	heniaio-
single	singuli-	monere-

six	sex	hek-
six	sex-	hex-, hexa-
six hundred	sescenti	hexakosi-
six hundredth	sescentensimi-	hexakosiosto-
six thousand	sex milia	hexakischilio-
six x	sexiens, -ies	hexaki-
six parts, into	sexfidi-	hexachei-, hexacha-
sixfold	sextuplici-	hexaplasio-, hexaple-,
		hexaploo-
sixteen	sedecim	hekkaideka-
sixteen x	sedeciens, -ies	hekkaidekaki-
sixteenth	sextus decimi-	hekkaidekato-
sixth	sexti-	hekto-
sixth	sexti-	hexa-
sixtieth	sexagensimi-	hexekosto-
sixty	sexaginta	hexekonta-, hexekontado-
sixty thousand	sexaginta milia	hexakismyrio-
sixty x	sexagiens, -ies	hexekontaki-
so many x, so often	totiens, -ies	tosaki-
solitary	solitarii-	monacho-
solitary	solitarii-	monadiko-
solitary	solitarii-	monia-
span	dodranti-	spithame-
ten	decem	deka-
ten parts, into	deciens, -ies	dekache
ten thousand	decem milia	dekachilio-
ten thousand	decem milia	myriado-
ten thousand x	deciens miliens, -ies	myriaki-
ten thousandfold	deciens millimodi-	myrioplasio-
ten thousandth	deciens millensimi-	myriadiko-
ten x	deciens, decies	dekaki-
tenfold	decemplici-	dekaplasio-
tenfold	decuplici-	dekaplasio-
tenth	decim-	dekato-
tertian	tertiani-	diatrito-
third	tertii-	trito-
thirteen	tredecim	treiskaideka-, triskaideka-
thirteen x	terdeciens, -ies	triskaidekaki-
thirteenfold	tredecimplici-	triskaidekaplasio-
thirteenth	tertius decimi-	triskaidekato-
thirtieth	tricensimi-	triakosto-
thirtieth part	tricensima pars	triakontamorio-,
		triakostamorio-
thirty	triginta	triakado-, triakonto-
thirty thousand	triginta milia	trismyrio-
thirty thousandfold	triginta millimodi-	trismyrioplasio-
thirty thousandth	triciens millensimi-	trismyriosto-
thirty x	triciens	triakontaki-
thirty-fifth	quinque et tricensimi-	triakostopempto-
thirty-second	alter et tricensimi-	triakostodyo-
thirty-three thousand	tres et triginta milia	trischiliotrismyrio-

thirtyfold	trigintuplici-	triakontaplasio-
thousand	mille-	chiliado-
thousand	mille-	chilio-
thousand & fiftyfold	millimodi et quinquagintuplici-	chiliokaipentekostaplasio-
thousand x	miliens, -ies	chiliaki-
thousandfold	millimodi-	chilioplasio-
thousandth	millensimi-	chiliosto-
three	tres, tri-	trei-
three	tres, tri-	tri-, triado-
three & a half	tres et dimidii-	triplasiephemiso-
three & a quarter	tres et quarta pars	triplasiepitetarto-
three & a third	tres et tertia pars	triplasiepitrito-
three & four-fifths	tres et quattuor partes	triplasiepitetramere-
three & one-fifth	tres et quinta pars	triplasiepipempto-
three & one-seventh	tres et septima pars	triplasiephebdomo-
three & three-quarters	tres et tres partes	triplasiepitrimere-
three & two-fifths, less by	tres et bis quinta pars	hypotriplasiepidipempto-
three & two-thirds	tres et duae partes	triplasiepidimere-
three hundred	trecenti	triakosio-
three hundredfold	trecentuplici-	triakosiochoo-
three thousand	tria milia	trischilio-
three thousandth	ter millensimi-	trischiliosto-
three ways	trivii-	diatricho-
three x, thrice	ter	tris-
three, in	trifarii-	tricha-, triche-, tricho-
three, the no.	ternii-	trittuo-, tritu-, trittua-
three-quarters	tres partes	trimoiriaio-
three-quarters	tres partes, dodrans	hypepitetarto-
threefold, triple	triplici-	triadiko-
threefold, triple	triplici-	trichthadio-
threefold, triple	triplici-	tridymo-, trizygo-
threefold, triple	triplici-	triphasio-, triphato-
threefold, triple	triplici-	triplaco-, triploci-, triploo-
threefold, triple	triplici-	triplasio-
threefold, triple	triplici-	trisso-, tritto-, trixo-
thrice, three x	ter	triaki-
thrice, three x	ter	trissaki-
trifid	trifidi-	trikro-
trifid, three-cleft	trifidi-	trischide-, trischisto-
tripartite	tripartiti-	trichtha-
tripartite	tripartiti-	trimere-, trimoiro-
twelfth	duodecim-	dodekato-
twelve	duodecim	dodeka-
twelve	duodecim	dyodeka-
twelve	duodecim	dyokaideka-
twelve thousand	duodecim milia	dodekakischilio-
twelve x	duodeciens, -ies	dodekaki-
twelvefold	duodecimplici-	dodekasemo-
twelvefold	duodecuplici-	dodekaplasio-
twenty	viginti	ikosi-, ikosa-
twenty thousand	viginti milia	dismyrio-

142

twenty x	viciens, -ies	ikosaki-
twenty-eightfold	duodetrigintaplici-	oktokaieikosaplasio-
twenty-fifth	quinque et vicensimi-	pentekaieikosto-
twenty-five	quinque et viginti	pentekaieikosi-
twenty-sevenfold	septem et vigintiplici-	heptakaieikosaplasio-
twenty-two	duo et viginti	ikosiduo-
twentyfold	vigintiplici-	ikosaplasio-
twice	duplici-	di-
twice, in two	bis	dicha-, diche-, dicho-
twice, in two	bis	dyasmo-
twin	gemini-	didymo-
two	bi-	amphi-
two	dua-, duo-	di-
two	duo	dyo-
two & a half	duo semis	diplasiephemiso-
two & four-fifths	duo et quattuor partes	diplasiepitetramero-
two & four-fifths	duo et quattuor partes	diplasiepitetrapempto-
two & half	duo et dimidii-	penthemimere-
two & one-fifth	duo et quinta pars	diplasiepipempto-
two & one-quarter	duo et quarta pars	diplasiepitetarto-
two & one-sixth	duo et sexta pars	diplasiepiekto-
two & one-third	duo et triens, -ies	diplasiepitrito-
two & three-quarters	duo et tres partes	diplasiepitrimero-
two & two-thirds	duo et duae partes	diplasiepidimero-
two & two-thirds	duo et duae partes	diplasiepidimoiro-
two & two-thirds	duo et duae partes	diplasiepiditrito-
two hundred	ducenti-	diakosia-
two hundred x	ducentiens, -ies	diakosiaki-
two hundred-fold	ducentuplici-	diakosiaplasio-
two thousand	duo milia	dischilio-
two, cut in	bisecti-	diakopto-
two, cut in	bisecti-	diatemno-
two, cut in	bisecti-	diatomo-
two, cut in	bisecti-	dimero-, -e
two-rowed	bifarii-	diphasio-
two-rowed	bifarii-	disticho-
two-thirds	duae partes, bes	hypepitrito-
		(fide L&S, non LSJ)
twofold	duplici-	disso-
uniform	uniformi-	monoide-
unity	uniti-	monoteto-
very few	perpauci-	oligosto-
very first, the	perprimi-	protisto-
zero x	nulli-	oudenaki-